THE LARGE PLANT-EATERS
The Encyclopedia of the Animal World:
Mammals

Managing Editor: Lionel Bender
Art Editor: Ben White
Designer: Malcolm Smythe
Text Editor: Miles Litvinoff
Assistant Editor: Madeleine Samuel
Project Editor: Graham Bateman
Production: Clive Sparling

Media conversion and typesetting:
Robert and Peter MacDonald

AN EQUINOX BOOK

Planned and produced by:
Equinox (Oxford) Limited,
Musterlin House, Jordan Hill Road,
Oxford OX2 8DP

Prepared by Lionheart Books

Library of Congress
Cataloging-in-Publication Data
Stidworthy, John, 1943-
Mammals: the large plant-eaters/John
Stidworthy.
p.96, cm.22.5×27.5 (The Encyclopedia of
the animal world)
Bibliography: p.1
Includes index.
Summary: Introduces mammals which live
and feed on the plains and savannahs, such
as gazelles, tapirs, and kangaroos.

1. Grassland fauna – Juvenile literature.
2. Mammals – Juvenile literature. 3.
Grazing – Juvenile literature. [1.
Grassland animals. 2. Mammals. 3.
Grazing.] I. Title. II. Series.

QL115.S75 1988   599-dc19
88-16935

ISBN 0-8160-1960-6

Published in North America by
Facts on File, Inc.,
460 Park Avenue South,
New York, N.Y. 10016

Origination by Alpha Reprographics Ltd,
Harefield, Middx, England

Printed in Italy.

10 9 8 7 6 5 4 3 2 1

FACT PANEL: Key to symbols denoting general features of animals

SYMBOLS WITH NO WORDS

Activity time

● Nocturnal

◗ Daytime

◒ Dawn/Dusk

○ All the time

Group size

◧ Solitary

▦ Pairs

◨ Small groups (up to 10)

■ Herds/Flocks

◪ Variable

Conservation status

☠ All species threatened

☠ Some species threatened

No species threatened (no symbol)

SYMBOLS NEXT TO HEADINGS

Habitat

◤ General

◢ Mountain/Moorland

◣ Desert

≈ Sea

Amphibious

Tundra

◹ Tundra

◿ Forest/Woodland

● Grassland

≈ Freshwater

Diet

■ Other animals

◻ Plants

◪ Animals and Plants

Breeding

◎ Seasonal (at fixed times)

○ Non-seasonal (at any time)

CONTENTS

PREFACE

The National Wildlife Federation

For the wildlife of the world, 1936 was a very big year. That's when the National Wildlife Federation formed to help conserve the millions of species of animals and plants that call Earth their home. In trying to do such an important job, the Federation has grown to be the largest conservation group of its kind.

Today, plants and animals face more dangers than ever before. As the human population grows and takes over more and more land, the wild places of the world disappear. As people produce more and more chemicals and cars and other products to make life better for themselves, the environment often becomes worse for wildlife.

But there is some good news. Many animals are better off today than when the National Wildlife Federation began. Alligators, wild turkeys, deer, wood ducks, and others are thriving – thanks to the hard work of everyone who cares about wildlife.

The Federation's number one job has always been education. We teach kids the wonders of nature through *Your Big Backyard* and *Ranger Rick* magazines and our annual National Wildlife Week celebration. We teach grown-ups the importance of a clean environment through *National Wildlife* and *International Wildlife* magazines. And we help teachers teach about wildlife with our environmental education activity series called *Naturescope*.

The National Wildlife Federation is nearly five million people, all working as one. We all know that by helping wildlife, we are also helping ourselves. Together we have helped pass laws that have cleaned up our air and water, protected endangered species, and left grand old forests standing tall.

You can help too. Every time you plant a bush that becomes a home to a butterfly, every time you help clean a lake or river of trash, every time you walk instead of asking for a ride in a car – you are part of the wildlife team.

You are also doing your part by learning all you can about the wildlife of the world. That's why the National Wildlife Federation is happy to help bring you this Encyclopedia. We hope you enjoy it.

Jay D. Hair, President
National Wildlife Federation

INTRODUCTION

The *Encyclopedia of the Animal World* surveys the main groups and species of animals alive today. Written by a team of specialists, it includes the most current information and the newest ideas on animal behavior and survival. The Encyclopedia looks at how the shape and form of an animal reflect its life-style – the ways in which a creature's size, color, feeding methods and defenses have all evolved in relationship to a particular diet, climate and habitat. Discussed also are the ways in which human activities often disrupt natural ecosystems and threaten the survival of many species.

In this Encyclopedia the animals are grouped on the basis of their body structure and their evolution from common ancestors. Thus, there are single volumes or groups of volumes on mammals, birds, reptiles and amphibians, fish, insects and so on. Within these major categories, the animals are grouped according to their feeding habits or general life-styles. Because there is so much information on the animals in two of these major categories, there are four volumes devoted to mammals (*The Small Plant-Eaters; The Hunters; The Large Plant-Eaters; Primates, Insect-Eaters and Baleen Whales*) and three to birds (*Waterbirds; Aerial Hunters and Flightless Birds; Plant- and Seed-Eaters*).

This volume, *Mammals – The Large Plant-Eaters*, includes entries on all of those *large* mammals that eat plants – elephants, sea cows, the panda, for example, as well as all of the hoofed mammals. Together they number some 270 species. Within this category are not only all the mammals that scientists refer to as the large herbivores (large plant-eaters such as deer, camels, goats, pigs, rhinos and zebras), but also the pouched mammals (marsupials) that have much in common with them, the Australian kangaroos and wallabies.

The animals in this group have two things in common. They are large and they are adapted to eating plants. They too are also a very diverse group of mammals. Besides including the largest land animals – the hippos, rhinos and elephants – they include a number of unusual marine animals, such as the dugongs and manatees. They also include mammals with antlers, such as deer and gazelle, and those with tusks, ones like elephants and wild pigs.

One of the animals included here is the Giant panda, a large animal that is closely related to the carnivores, in particular bears, but that feeds mostly on bamboo shoots.

Each article in this Encyclopedia is devoted to an individual species or group of closely related species. The text starts with a short scene-setting story that highlights one or more of the animal's unique features. It then continues with details of the most interesting aspects of the animal's physical features and abilities, diet and feeding behavior, and general life-style. It also covers conservation and the animal's relationships with people.

A fact panel provides easy reference to the main features of distribution (natural, not introductions to other areas by humans), habitat, diet, size, color, pregnancy and birth, and lifespan. (An explanation of the color coded symbols is given on page 2 of the book.) The panel also includes a list of the common and scientific (Latin) names of species mentioned in the main text and photo captions. For species illustrated in major artwork panels but not described elsewhere, the names are given in the caption accompanying the artwork. In such illustrations, all animals are shown to scale; actual dimensions may be found in the text. To help the reader appreciate the size of the animals, in the upper right part of the page at the beginning of an article are scale drawings comparing the size of the species with that of a human being (or of a human foot).

Many species of animal are threatened with extinction as a result of human activities. In this Encyclopedia the following terms are used to show the status of a species as defined by the International Union for the Conservation of Nature and Natural Resources:

Endangered – in danger of extinction unless their habitat is no longer destroyed and they are not hunted by people.

Vulnerable – likely to become endangered in the near future.

Rare – exist in small numbers but neither endangered nor vulnerable at present.

A glossary provides definitions of technical terms used in the book. A common name and scientific (Latin) name index provide easy access to text and illustrations.

ELEPHANTS

In the blazing red of an African sunset, a herd of elephants makes its way down to the river bank. In the middle of the herd, close to their mothers, are two young babies. Suddenly the leader stops. She puts up her trunk and fans out her ears. She senses danger. The elephants halt and bunch up. From some thorn bushes two lionesses appear. The leading elephant gives a squeal of rage, tucks up her trunk and charges. The lions scatter. No other animal stays in the path of a charging elephant.

ELEPHANTS Elephantidae
(2 species)

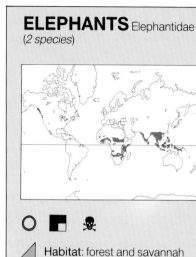

○ ■ ☠

◢ **Habitat:** forest and savannah grassland.

■ **Diet:** grasses, shrubs and trees, including twigs, leaves and bark.

◎ **Breeding:** usually during the annual wet season. 1 young after pregnancy of 22 months.

Size: to 11ft in height, 13,200lb in weight. Females shorter, to 8,800lb.

Color: gray-black, sometimes with pinkish patches. Often appears red-brown from dust thrown over body.

Lifespan: 60 years.

Species mentioned in text:
African elephant (*Loxodonta africana*)
Asian elephant (*Elephas maximus*)

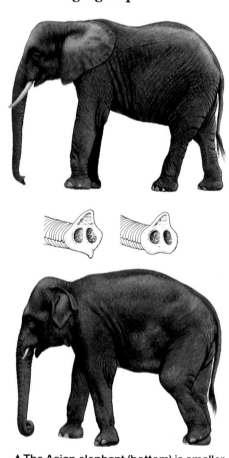

▲ The Asian elephant (bottom) is smaller than the African and has smaller ears. It has a single lip, not two, at the trunk tip.

► Elephants have a good sense of smell. Trunks are held high to catch scents.

The African elephant is the biggest living land animal. The slightly smaller Asian species is rare now as a wild animal, and only about 50,000 are left.

GIANTS OF THE LAND

The biggest African elephant ever measured stood 13ft tall, about as high as a cottage. It was a male (bull) elephant. These often reach 10½ft tall, and nearly 13,200lb in weight. Females (cows) are shorter, and weigh up to 8,800lb. The biggest African elephants are those from savannah country. Those from forests near the equator are usually smaller, and have rounder ears.

Since elephants are so big, they have problems keeping cool. Their ears give a large surface from which heat can be lost. The elephants also use them as fans.

GIANT APPETITES

Elephants have an appetite to go with their size. An adult needs about 330lb of plant food a day. This is equal to about six small bales of hay. But much of the food is of poor quality. Bacteria in the gut help the elephant break down its food. Even so, nearly half passes through the elephant without being digested. Elephants have to spend about 16 hours out of every 24 feeding. Water too is drunk in large amounts. An elephant needs about 18 gallons a day (equal to 140 pint cartons of milk).

TRUNK CALLS

An elephant gathers all its food and water with its trunk. This is a combined nose and upper lip. The trunk is full of muscles which can move it in all directions with precision or strength. It can be used to pick a single leaf or berry, using the "lips" at the tip. Or instead, it may be used to rip a branch from a tree before lifting it to the mouth. Water is always sucked into the trunk, then squirted into the mouth and swallowed. Another use for the trunk is as a snorkel when bathing.

▲Where their numbers are high, elephants can cause enormous changes to an area by uprooting trees to feed on their tops.

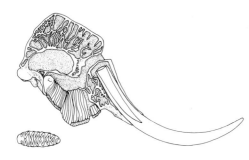

▲In the elephant's huge skull 24 chewing teeth grow in sequence. Of these only four are used at a time. When worn, they are lost and replaced from behind.

The trunk is also important in making sounds, and helping elephants to "keep in touch". Elephants greet one another by putting the trunk tip to the other's mouth. Mothers reassure their babies (calves) by touching and guiding them with the trunk. Sniffing and touching with the trunk tells an elephant much about its surroundings.

LIVING TOGETHER

Adult bull elephants live alone, or sometimes in small groups. They join the females only for mating. Cow elephants live in herds accompanied by their calves. Bulls leave the herd as they become adult, but cows stay with their mother. The herd is usually led by the oldest and largest cow, the matriarch. She, by example, shows when and where to move. She also decides how to react to threats. She

may charge an enemy, or lead the herd away. An elephant's lifespan is nearly as long as a human's. The leader has had time to learn many useful things. By following her behavior, other cows in the herd can gain experience for the time when they may have to lead a herd. The matriarch is often too old to produce young.

BIG BABIES

The elephant has the longest pregnancy of all mammals. At birth, the newborn baby weighs 265lb – more than most adult humans. It sucks milk from its mother using its mouth, not its trunk. In the first months the trunk is almost useless. The calf takes milk until about 2 years old, but eats plants after only a few months.

Elephants grow fast until they are about 15 years old. After this growth

In Asia people have used elephants for thousands of years. A few are still used for work in some timber forests. They can move logs over 2,200lb in weight and can work where tractors cannot.

The elephant's most important teeth are the huge grinding cheek teeth. Without these it cannot eat properly. An old elephant which has worn down its last teeth will starve.

ELEPHANTS AND PEOPLE
Although elephants are so big, people can tame and train them. Working elephants are usually females, as these are better tempered than males. People have used Asian elephants much more than African ones for working, but the African general Hannibal crossed the Alps with 57 African war elephants on his way to attack Rome in 218 BC. Tame elephants, often painted or with bright costumes, take part in some Asian festivals. In some places light-colored ("white") elephants are especially respected.

HEADING FOR EXTINCTION?
Hunting for ivory to make ornaments is perhaps the greatest danger facing elephants. Each year, about 100,000 African elephants are killed, most of them illegally, for their tusks. Also, as human numbers increase, farms are built where elephants used to roam freely. When crammed together, elephants destroy habitats. Occasionally they must be culled. In Africa there may still be a million elephants, but their numbers are falling fast.

may slow down, but elephants, unlike many mammals, go on growing throughout life. Their tusks too continually grow.

WORLD'S BIGGEST TEETH
The elephant's tusks are simply enormous front (incisor) teeth. The tough material of which they are made is known as ivory. The longest tusk on record was 11½ft long. The heaviest pair known weighed a total of 464lb. Most are much smaller, but, even so, a big bull elephant carries a great weight in tusks. Cows have smaller, more slender, tusks. In Asian elephants the tusks of cows are so small they hardly stick out beyond the lips. The tusks are occasionally used in feeding, or may be used to threaten or fight a rival. But most of the time they are not used at all.

▲Bull elephants sometimes fight one another for the chance of mating with a cow that is on heat.

Fighting consists of charges and shoving matches. Sometimes the trunk and tusks are used in wrestling.

Usually the smaller elephant gives way once it has tested its strength. Only rarely is one of the animals hurt.

HORSES AND ASSES

A herd of wild horses moves slowly forwards, heads down, cropping grass. The only adult male lifts his head, alerted by the arrival of a male horse from another herd. He walks to challenge it. The stranger hesitates. The herd master drops his head and charges, baring his teeth as if to bite. The stranger turns and flees. The herd goes back to grazing.

HORSES AND ASSES
Equidae (*4 species*)

 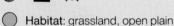

Habitat: grassland, open plain and desert.

Diet: mainly grasses, some bark, leaves and fruits.

Breeding: 1 or 2 foals after pregnancy of 11½ months, at the season when plant growth is best. Females mate soon after giving birth.

Size: head-body 6½-7ft plus 20in tail; weight 600-770lb.

Color: yellowish-brown to gray, lighter below. Tail and mane darker, often a dark back stripe.

Lifespan: 25 years.

Species mentioned in text:
African ass (*Equus africanus*)
Asiatic ass (*E. hemionus*)
Domestic horse (*E. caballus*)
Przewalski's horse (*E. przewalskii*)

There are three species of wild horses and asses plus the Domestic horse, of which there are many breeds. Wild horses live mostly on open plains in dry regions. Przewalski's horse is native to Mongolia. It is probably extinct in the wild. None have been seen there since 1968. Another form of the same species, the Tarpan, used to live on the European steppes. These wild horses were the ancestors of our domestic horses.

The Asiatic ass is a strongly built animal with broad hoofs that lives in deserts. There are four slightly different races, all existing in small numbers. The African wild ass roams rocky deserts in North Africa. It is the smallest of the horse family, and is the ancestor of the domestic donkey.

Asses withstand water loss well, and can journey far to find sparse supplies of food and water. But even in deserts they are not always safe from humans, and numbers are declining.

LIVING IN THE OPEN
Horses and asses are built for eating grass and running fast to escape danger. They gather grass with their lips and the top and bottom front teeth (incisors). Their chewing teeth and gut can cope with tough grasses.

Horses have long legs, each with a single hoof, or toe-nail. The legs are light and easy to swing when running. A wild horse can run at speeds of 30mph, and has plenty of stamina.

Horses have keen eyes. These are set far back on the head, and they can see danger from all directions. Their hearing is excellent, and the sense of smell is good. Voice plays a part in social behavior, and so does scent. Horses indicate moods to one another by changing ear, mouth and tail positions.

CHANGED BY HUMANS
Over the years people have bred wild horses for particular jobs. There are now breeds as different as carthorses, taller than a man, and Shetland ponies, only waist-high.

Domestic horses have a mane that falls sideways, unlike that of any wild species. The big herds of "wild" horses seen in America and Australia are descended from tame ones allowed to run loose.

▶**Wild species of the horse family**
A Przewalski's horse **(1)** (male) charges
a rival. An African ass **(2)** lashes out with
her feet. An Asiatic ass marks territory
with a dung-heap **(3)** and curls its lip as it
sniffs a female **(4)**.

▼No modern species of horse lived in
America until introduced by Europeans
in the 16th century. Then, allowed to go
wild, they became the mustangs used by
Plains Indians and cowboys.

ZEBRAS

A long column of zebras crosses the dry plain, each animal following the ones in front. They are in search of water. Suddenly a leopard drops to the ground from the branch of a tree, just missing a foal that is lagging behind its mother. There is a moment of panic. Zebras dash in circles, creating a confusing mass of black and white. The foal escapes. As the disappointed leopard stalks off, calm returns. The zebras plod on towards the next river.

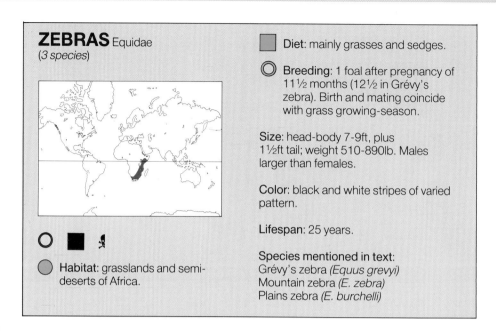

ZEBRAS Equidae
(*3 species*)

○ ■ ☠

🔵 Habitat: grasslands and semi-deserts of Africa.

▢ Diet: mainly grasses and sedges.

◎ Breeding: 1 foal after pregnancy of 11½ months (12½ in Grévy's zebra). Birth and mating coincide with grass growing-season.

Size: head-body 7-9ft, plus 1½ft tail; weight 510-890lb. Males larger than females.

Color: black and white stripes of varied pattern.

Lifespan: 25 years.

Species mentioned in text:
Grévy's zebra (*Equus grevyi*)
Mountain zebra (*E. zebra*)
Plains zebra (*E. burchelli*)

A zebra is a striped horse. There are three species of zebra. Apart from stripes, and the fact that they all live in Africa, the species have little in common, and are only related genetically.

WHY BE STRIPED?

There are various ideas about why it is useful to a zebra to have stripes. One theory is that the stripes are mainly for use among zebras, as a bright signal that allows them to follow one another's movements. Zebras certainly seem to be attracted by stripes. Perhaps at times there are other uses for stripes, such as camouflage.

MIGRATING MULTITUDES

The most numerous of the zebras is the Plains zebra, found from South to East Africa. In many areas its broad black stripes and rather dumpy body are a familiar sight. The Plains zebras of some areas, such as the Serengeti and Botswana, make long migrations to make use of seasonal growth of grass. At this time many thousands of zebras may move together, but for most of the year the typical herd is a

▶A herd of Plains zebras drinks at a waterhole in Etosha National Park in Namibia.

◄Each zebra has a unique pattern of stripes. This may help them recognize one another. This zebra is unusually marked.

male (stallion) and his group of females (mares) and their young (foals).

THREATENED ZEBRAS

The Mountain zebra is found in mountain grasslands of south-west Africa. It has narrower stripes than the Plains zebra and a fold of skin, the dewlap, under the throat. It has a "grid-iron" arrangement of stripes over the rump. It lives in small herds. Populations are small. They are protected in national parks, but this species is listed as vulnerable.

Grévy's zebra lives in northern East Africa. It is the largest zebra, and has the narrowest stripes. It has a tall mane and big rounded ears. It lives in small herds in thorn scrub country. Once fairly common, it has been hunted for its beautiful coat in recent years. It is now endangered.

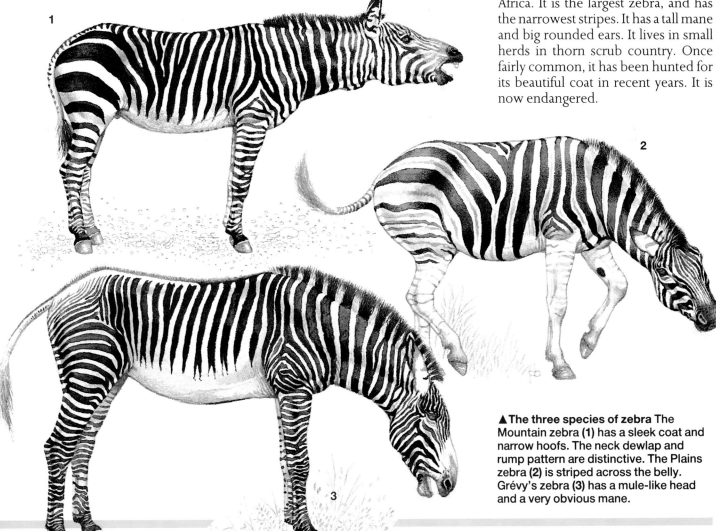

▲The three species of zebra The Mountain zebra (1) has a sleek coat and narrow hoofs. The neck dewlap and rump pattern are distinctive. The Plains zebra (2) is striped across the belly. Grévy's zebra (3) has a mule-like head and a very obvious mane.

TAPIRS

A nose appears out of the jungle river and cautiously sniffs the air. Satisfied that the scent which alarmed it has gone, the nose's owner hauls itself out of the water and on to the bank, to resume the night's feeding. Sniffing and feeling with its flexible nose, the tapir chooses the choicest morsels from the bushes as it passes. It forces its bulky body between the bushes, then walks along the same well-worn track that it took to the water.

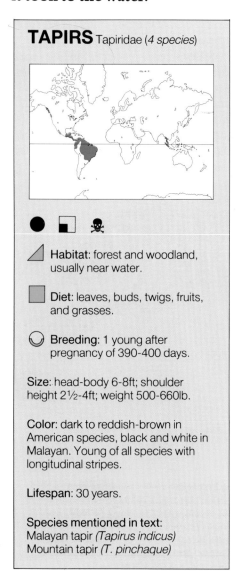

TAPIRS Tapiridae (4 species)

● ■ ☠

◢ Habitat: forest and woodland, usually near water.

▨ Diet: leaves, buds, twigs, fruits, and grasses.

◖ Breeding: 1 young after pregnancy of 390-400 days.

Size: head-body 6-8ft; shoulder height 2½-4ft; weight 500-660lb.

Color: dark to reddish-brown in American species, black and white in Malayan. Young of all species with longitudinal stripes.

Lifespan: 30 years.

Species mentioned in text:
Malayan tapir (Tapirus indicus)
Mountain tapir (T. pinchaque)

Tapirs live in two quite separate parts of the world. One species, the Malayan tapir, lives in South-east Asia. The other three live in Central and South America. Most tapirs inhabit lowland areas. The Mountain tapir inhabits mountain forests, and may wander above the tree-line.

SEARCHING SNOUT

Tapirs have a compact shape, ideal for pushing through dense jungle under-

▲ Tapirs like water, and will also wallow in mud.

growth. The snout has a short, fleshy trunk. This is a combined nose and upper lip, and has nostrils at the tip. Tapirs use their trunk as a sensitive "finger" to pull leaves and shoots to the mouth. They eat many kinds of plant food, but prefer to browse on green shoots in forest clearings and along river banks.

3

Tapir tracks are often the only sign that the animals are around. They show three hoofed toes, although there is a small fourth toe on the front feet. Like rhinos and horses, tapirs take the weight down the middle toe of the feet. The side toes are smaller.

GUIDED BY SMELL

Tapirs are most active at night. They have large ears, and their hearing is good, but not as acute as the sense of smell. Their eyes are small and provide only reasonable vision. Tapirs walk with their nose to the ground. This helps them to sniff out where they are, and to detect the scent of other tapirs or predators. They follow familiar routes, and mark these with their urine.

Apart from mothers with young, tapirs usually live alone. They may range widely. They are good swimmers, and will take to water and submerge to avoid danger. Disturbed away from water, tapirs crash off into the bush, or defend themselves by biting. Mating starts with a noisy courtship. The pair of tapirs circles at speed and both animals squeal. They nip and prod each other. At the end of pregnancy, the female finds a secluded lair in which to give birth. Tapirs are fully adult at about 3 years old.

UNCERTAIN FUTURE

Tapirs are hunted for food, sport and for their thick skins, which give good leather. But the biggest threat to them is the loss of their habitat. Soon they may only survive in areas specially set aside for wildlife.

◀The species of tapir The Mountain tapir (1) has a woolly coat. The Brazilian tapir (*Tapirus terrestris*) (2) and Baird's tapir (*T. bairdi*) (3) have bristly manes. The black and white Malayan tapir (4) is hard to see at night. Baby tapirs lose their stripes by 6 months old.

RHINOCEROSES

An African Black rhinoceros wallows in a muddy pool, her tiny calf beside her. Suddenly sounds of movement come from a thicket not far away. The mother rhino stands up and charges towards the thicket. An animal scurries away. She stops, turns, and trots back to the calf.

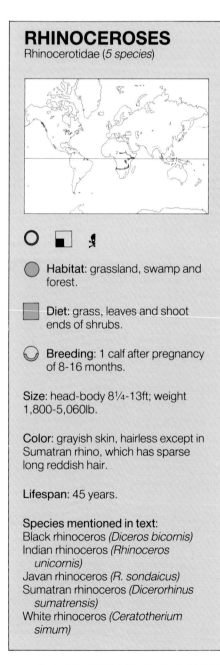

RHINOCEROSES
Rhinocerotidae (5 species)

○　■　🐾

● **Habitat:** grassland, swamp and forest.

■ **Diet:** grass, leaves and shoot ends of shrubs.

◑ **Breeding:** 1 calf after pregnancy of 8-16 months.

Size: head-body 8¼-13ft; weight 1,800-5,060lb.

Color: grayish skin, hairless except in Sumatran rhino, which has sparse long reddish hair.

Lifespan: 45 years.

Species mentioned in text:
Black rhinoceros (Diceros bicornis)
Indian rhinoceros (Rhinoceros unicornis)
Javan rhinoceros (R. sondaicus)
Sumatran rhinoceros (Dicerorhinus sumatrensis)
White rhinoceros (Ceratotherium simum)

There are five species of rhinoceros. The Black rhino and the White rhino both live in Africa. They differ little in color, and both have two horns, but they can be told apart by the shape of their faces. The Black rhino is a browsing animal, feeding on leaves from bushes. It has a long pointed upper lip which helps it pull food into its mouth. The White rhino is a grazer, and has a wide muzzle. It crops many blades of grass at once.

The other three species of rhino live in Asia. The Indian rhino has an armor-plated look, with big folds of skin above the legs. This, and the African White rhino, are the two biggest species. Males grow to 6ft tall and 5,060lb in weight, and females to 5½ft and 3,750lb. The Javan rhino is smaller and has less obvious body folds. The Indian and the Javan rhino have only one horn.

The smallest rhino is the Sumatran, which has two horns, and has a thin coat of reddish hair. Javan and Sumatran rhinos browse from bushes and saplings. Indian rhinos pull in shrubs and tall grass with the upper lip, but can fold the tip away to graze short grass.

HORNS AND HAIR
Rhino horns are unusual in that they lie along the middle of the snout and, unlike the horns of sheep, cattle and antelope, they do not have a bony center. They are not firmly attached to the skull. Rhino horns grow from the skin, and are made of the same chemical as hair and claws. They are hard and solid, but are made up of many fibres. African rhinos sometimes have front horns 5ft in length.

Except in the Sumatran rhino, hair is only visible as eyelashes, ear fringes and tail tassels. These animals do not need fur to keep warm.

SENSES
Rhinos rely mostly on their sense of smell to explore their surroundings.

▲Rhino courtship and mating may take several hours. Courtship is rough, with long chases and "fights" – sparring with the horns – between male and female.

They have rather poor vision. They cannot pick out a person standing still more than 100ft away. Their hearing is good, and they turn their tubular ear flaps towards sounds in which they are interested. Rhinos can make many sounds, from roars and squeals to bleats which sound too gentle for such massive animals.

Partly because of their poor vision, perhaps, some species are apt to make sudden charges at intruders. The Black rhino has a reputation for aggression, and the Indian rhino too may make apparently unprovoked attacks. Most rhino charges are not carried through. African rhinos attack with their horns, but the Asian species may bite a supposed enemy. These are the same methods that male rhinos use when fighting each other.

A SOLITARY LIFE
Most rhinos live alone. A calf may stay with its mother for 2 or 3 years. Sometimes several rhinos are found together around a good feeding site. But most rhinos prefer to be by themselves.

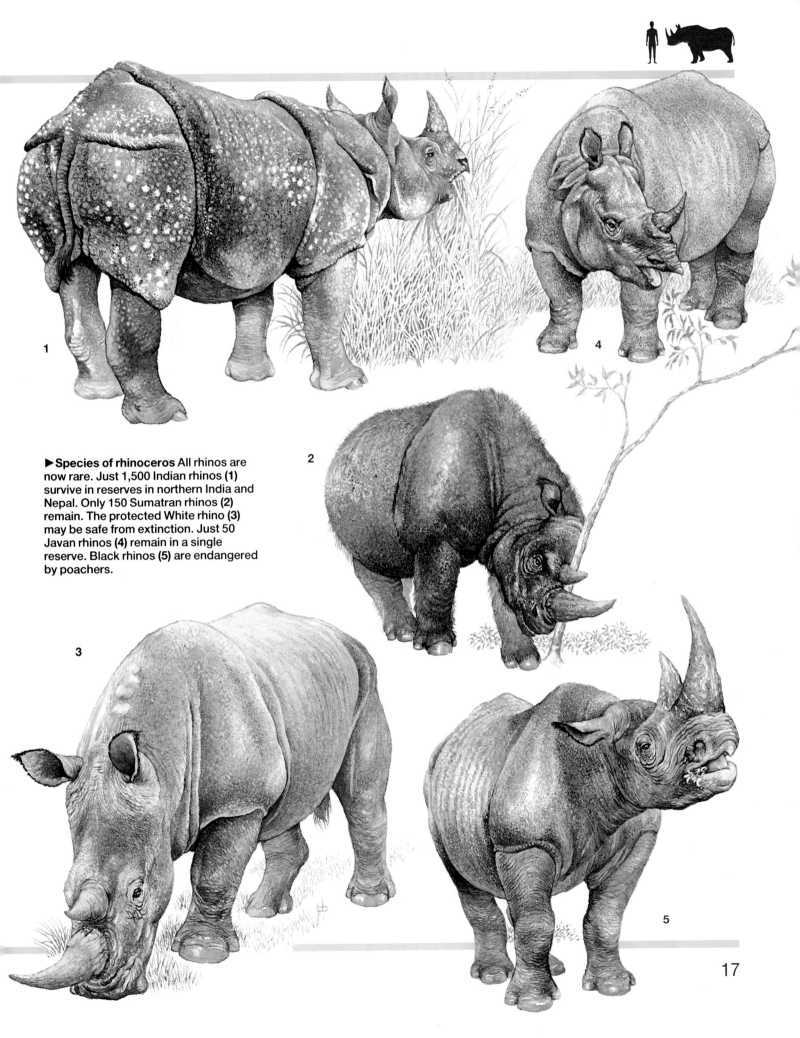

▶**Species of rhinoceros** All rhinos are now rare. Just 1,500 Indian rhinos **(1)** survive in reserves in northern India and Nepal. Only 150 Sumatran rhinos **(2)** remain. The protected White rhino **(3)** may be safe from extinction. Just 50 Javan rhinos **(4)** remain in a single reserve. Black rhinos **(5)** are endangered by poachers.

The exception is the White rhino, which may form small herds of six or seven animals. Even so, adult males are usually solitary.

Adult males often claim an area for themselves, keeping out other males. These may be repelled by ritual fighting, such as sparring with horns, or wiping the horns on the ground. Sometimes a real battle begins, and the animals wound one another. Usually, the rituals are enough to keep the males spread apart. They also mark the edges of their territories with special piles of dung and urine.

LIFE HISTORY

Rhinos have long lives. They may be able to breed at 5 years old, but females bear a calf only every 2 or 3 years. Males may not be able to defend an area and breed until they are about 10 years old. Baby rhinos are small compared to their parents, weighing only about 90lb in the case of the Indian rhino, but they are still as heavy as most 12-year-old children. Rhino mothers usually find a quiet spot in which to give birth. They may leave tiny babies hidden while they feed elsewhere, but after a few days most baby rhinos move with their mother. Baby Indian and White rhinos tend to run in front of the mother, baby Black rhinos usually run behind.

SURVIVAL IN QUESTION

Some 40 million year ago, rhinos of various kinds were abundant in most warm regions of the world. Now these animals are in danger of extinction. They are hunted for their horn, which is believed to have medicinal properties. It is also used to make handles for daggers which are status symbols in parts of Arabia.

◄African White rhinos are usually peaceable and rather timid, in spite of their size. These grass-eaters are more often found in groups than other rhinos.

WILD PIGS

It is night in an African village. Once the people are asleep a family of bushpigs comes out of the forest into the villagers' small fields. Busily, they dig with their noses. They uproot and eat plants, and feed on small animals they disturb. By morning the pigs are gone, but patches in the fields look as if they had just been ploughed.

Three species of wild pig live in Africa. Four more live on islands in Southeast Asia. The smallest wild pig, the rare Pygmy hog, lives only in the tall grass of the Himalayan foothills. The most widespread wild pig is the Wild boar, found from Europe to eastern Asia. This is the species from which domestic pigs have been bred.

KNOWLEDGEABLE NOSES

All pigs have a large head and short neck. Their snout is very long, with a nose supported by a special movable bone. The end of the nose is round and flat like a disc. For pigs, the nose is a very important organ. It provides the

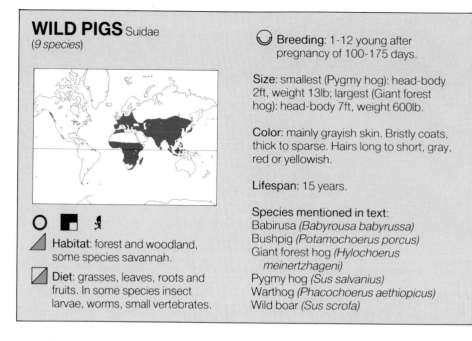

WILD PIGS Suidae
(9 species)

⬤ **Breeding**: 1-12 young after pregnancy of 100-175 days.

Size: smallest (Pygmy hog): head-body 2ft, weight 13lb; largest (Giant forest hog): head-body 7ft, weight 600lb.

Color: mainly grayish skin. Bristly coats, thick to sparse. Hairs long to short, gray, red or yellowish.

Lifespan: 15 years.

Species mentioned in text:
Babirusa *(Babyrousa babyrussa)*
Bushpig *(Potamochoerus porcus)*
Giant forest hog *(Hylochoerus meinertzhageni)*
Pygmy hog *(Sus salvanius)*
Warthog *(Phacochoerus aethiopicus)*
Wild boar *(Sus scrofa)*

○ ◧ ☠

◢ **Habitat**: forest and woodland, some species savannah.

◢ **Diet**: grasses, leaves, roots and fruits. In some species insect larvae, worms, small vertebrates.

animals with a keen sense of smell, and a good sense of touch at the snout tip. The nose is also used to dig in earth in search of food. A large part of a pig's brain is devoted to the movement and sensitivity of its nose.

Pigs have a coat of coarse bristles and a tasseled tail which they use to swat flies and to signal moods to one

another. They have four toes on each foot. The outside two toes are small and do not usually touch the ground. Some wild pigs feed almost entirely on plants, but other members of the family eat small animals too.

CURVING CANINES

The tusks of wild pigs are big canine teeth. Both the lower and the upper tusks curve upwards and outwards. The tusks can be deadly weapons, and Wild boar especially will fight back hard if threatened. They will fight off

▼ Species of wild pig fight in ways that reflect the shape and form of their weapons and armour. The Giant forest hog **(1)** fights head-on. Warthogs **(2)** and bushpigs **(3)** have warts to protect the face. Wild boar **(4)** slash at opponents' shoulders, where they have tough skin and hair.

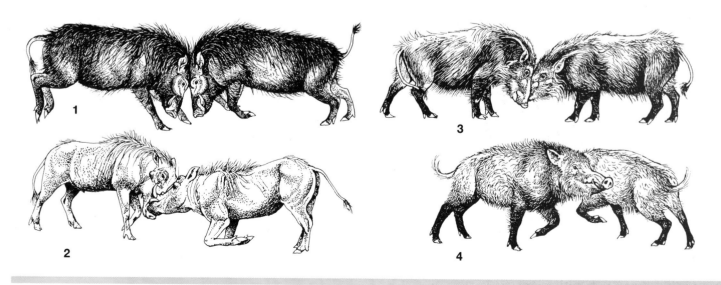

▲Although a Giant forest hog's upper tusks are more impressive, it is the smaller lower tusks that are the animal's main weapons.

▲A family of warthogs drinks at a pool. These African pigs often drink and feed on grass shoots and roots resting on their front "knees".

▼The Wild boar is one of the largest pigs. The shoulders are padded for protection in fights. In front of each eye is a slit containing a scent gland.

leopards or tigers. Warthogs also have big tusks, but they fight only as a last resort. Warthogs run from trouble with their tail held high like a flag. They will also take cover in burrows dug by other animals such as aardvarks.

The biggest tusks of all belong to the babirusa of Sulawesi and nearby Indonesian islands. This wild pig has upper tusks which grow straight up through the skin of the snout and then curve backwards. Sometimes they grow right over the snout to touch the forehead. The lower tusks are long too. They seem to be useless as weapons, although they may help males impress one another when competing for females. In all pigs males have bigger tusks than females.

Several kinds of pig have large warts on their face. These, and other toughened parts of the head, give some protection when they are fighting one another.

FAMILY GROUPS

Pigs go about in groups, called sounders, made up of a mother (sow) and her young (piglets). Young of most species have a striped coat. Sometimes females that are old enough to breed stay with their mother, so the group is bigger. Adult males (boars) often live alone, but groups are sometimes found. The strongest males in an area mate with the females. Courtship involves communication using sounds, such as squeaks and grunts, and scents. Some pigs have special scent-producing glands in front of the eyes, on the lips, or on the feet. They use them to mark their home areas.

PECCARIES

Day breaks over a Central American forest. A group of peccaries begin to stir. They have been resting in the shelter of an overhanging rock. One by one they get up and stretch. They nuzzle and groom one another. Two stand side by side, rubbing their faces against the other's body. They are marking one another with scent. This helps the animals recognize one another.

Peccaries are medium sized pig-like animals with long slender legs. There are three species. One was only recognized by scientists as late as 1975.

Peccaries eat all types of food. They feed mainly on roots, seeds and fruits. In drier areas many eat cacti. They also eat insect grubs and other, but small, animals. They have been known to eat snakes, including rattlesnakes. Their jaws have a strong up and down action. They can crush seeds and chop up bulbs and other roots.

AMERICAN PIGS

Peccaries are found only in the Americas, from southern USA to Argentina. There they are the equivalent of the true pigs that are found only in Europe, Asia and Africa. Peccaries are similar to pigs in their general shape, in their coarse bristly coats, and in the shape of the nose. They, too, can use the nose for rooting. They differ from pigs in their teeth. Peccaries have short sharp tusks, and the upper ones grow down, not turned upwards as in pigs. The peccary stomach is also more complicated. It has compartments in which bacteria help to break down plant food. But peccaries do not chew the cud like cattle.

Another feature of peccaries is the musk gland on the rump. The gland lies about 8in in front of the short tail. When a peccary is excited or annoyed the back hair and bristly mane stand on end. At the same time the musk gland gives off an unpleasant smell.

WANDERING HERDS

Peccaries are social animals and are rarely found singly. The Collared peccary lives in family groups of male, females and young, up to about 15 animals in all. These often combine with three or four similar groups to form large herds. The White-lipped peccary may form even larger herds, sometimes over 100 strong. The newly discovered Chacoan peccary is,

▲A young Collared peccary is cared for by its parents and all the other members of their group. It feeds on its mother's milk for just 2 months, but stays close to her for 6 months.

▶The Collared peccary lives in many types of country from tropical forest to dry scrubland. The spear-like tusks give the animal its alternative name javelina, from the Spanish for spear.

by peccary standards, a loner. Its groups number from 2 to 10.

NOISY GNASHING

As well as the scent messages that pass between them, peccaries can "talk" to one another with a variety of sounds. The alarm call is a short "woof". Lost babies cluck shrilly until their mother finds them. Angry peccaries gnash their teeth, making a rasping sound. If they fight, they may make a "laughing" cry. A coughing call brings a group together if it has spread out.

A group of peccaries lives within a fixed area. Where food is plentiful such an area may be only about 74 acres. (An acre is about the area of a full-sized football pitch). Where food is scarce a group may have a territory of up to 690 acres. Although peccaries are friendly with animals in their own group, they will fight unfamiliar peccaries.

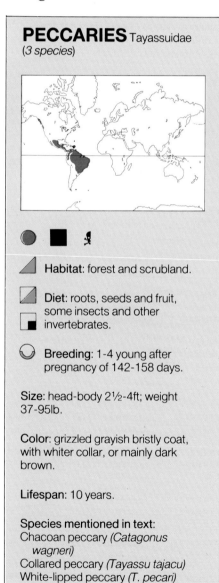

PECCARIES Tayassuidae
(*3 species*)

🔴 ⬛ ☠

🔺 Habitat: forest and scrubland.

◱ Diet: roots, seeds and fruit, some insects and other invertebrates.

☽ Breeding: 1-4 young after pregnancy of 142-158 days.

Size: head-body 2½-4ft; weight 37-95lb.

Color: grizzled grayish bristly coat, with whiter collar, or mainly dark brown.

Lifespan: 10 years.

Species mentioned in text:
Chacoan peccary (*Catagonus wagneri*)
Collared peccary (*Tayassu tajacu*)
White-lipped peccary (*T. pecari*)

▲ Peccaries use scent marking to "keep in touch". Individuals rub their scent glands on each other's coat, creating a communal scent.

DEFENDING THE GROUP

Both jaguars and pumas will catch and eat peccaries if they get the chance. So too will other predators such as the anaconda snake. Because peccaries live in large groups, they are not easy to take by surprise. If a herd is attacked, the members may give the alarm call and scatter, so confusing the attacker. Another tactic sometimes used is for a single peccary to remain behind and move towards the attacker. This may divert attention from the main herd, but often results in the capture of the lone animal.

Unselfish behavior seems common in these animals. Any adult, not just the parents, will allow a baby peccary to shelter from danger between their legs. The young are allowed to eat the best foods, and are not pushed out of the way by the adults.

A peccary litter consists of up to four babies. They are often born in a den, but can run after a few hours and soon begin to follow their mother. They eat solids from 1 month old.

HIPPOPOTAMUSES

In a deep pool in an African river the fish swim lazily. A small crocodile drifts by. Suddenly a huge animal springs across the bottom of the pool in a slow motion gallop, lightly touching the bottom with its toes. Then it floats to the surface, where just its eyes, nose and ears peek above water. It is a hippopotamus. On land the animal is heavy and cumbersome. Buoyed up by the water it is almost weightless and moves easily and gracefully.

The most familiar hippo is the larger of the two species. It is found in slow-moving rivers and lakes. It lives over much of Africa south of the Sahara where water and grassland are close to one another. The second species is the Pygmy hippo found in the forests of West Africa. This is a much rarer animal. It is tiny compared to its grassland cousin, and much more difficult to observe. Few details are known of its habits.

The common hippo has eyes, ears and nose at the very top of the head, so it can hear, see and breathe in air while mostly submerged. The nostrils and ears can be closed when under water. In the Pygmy hippo the eyes are further to the side of the head.

DAY-TIME LAZY-BONES
The common hippo spends most of the day in water. It seems to prefer slow-moving rivers, but also lives in lakes, and sometimes in estuaries. This aquatic habit takes the weight off its feet, and also prevents water loss through the skin. Water escapes through hippo skin four times as fast as through a person's in dry air. Hippo skin is smooth, with few hairs. Glands in the skin secrete a pink fluid which protects the skin from sunburn. Beneath the skin is a layer of fat 2in

HIPPOPOTAMUSES
Hippopotamidae (*2 species*)

Size: Pygmy hippopotamus: head-body 5ft; weight 400lb. Hippopotamus: head-body 11½ft; weight 7,040lb. Males larger than females.

Color: blackish gray, lighter or pinkish below.

Lifespan: 45 years.

Species mentioned in text:
Hippopotamus (*Hippopotamus amphibius*)
Pygmy hippopotamus (*Choeropsis liberiensis*)

● ■ ☠

Habitat: lakes, wallows and rivers during day, grassland at night. Pygmy hippo in swamp forest.

Diet: mostly land vegetation, especially short grasses.

◎ Breeding: 1 young after pregnancy of 190-240 days.

▶The hippo's lower tusks are up to 20in long. They are used in fights between rival males. Adult males are often scarred.

▶Hippos feed on land but often get rid of waste food (defecate) in water, so adding fertilizer to the water. Some kinds of fish stay close to hippos, either to graze on tiny plants that grow on the hippos skin, or to feed on the dung.

thick, which keeps the hippo warm in cool water.

Day is a lazy time for the hippo. It rests, swims a little, submerges for up to 5 minutes, and yawns repeatedly.

NIGHT-TIME GRAZERS
At night the hippo emerges from the water. This is when it does all its feeding. For 6 hours or so, the hippo crops grass with its huge lips. Where possible it feeds on patches of short grasses, or hippo lawns, close to

water, but sometimes has to travel several miles to find food. Hippos sometimes make long overland journeys to new bodies of water.

The strongest male in an area mates with all the females in it. He keeps rivals away by displays of yawning to show the teeth, by charges, and by loud grunting calls. As a last resort he will fight. Another way in which males stay spread apart is by marking the edge of their territory with dung. As the animal defecates it wags its short flat tail. This scatters the dung and makes an effective scent mark.

Sometimes males are solitary, but most hippos live in groups of 10 to 15. Most hippo babies are born in the rainy season, when the grasses are growing well. Pygmy hippos live alone, or in pairs.

CAMELS AND LLAMAS

A group of six camels is huddled in the middle of the desert. It is early winter in Central Asia. The wind is blowing across the vast empty plain. There are flurries of snow. The camels' thick woolly coats are caked with snow that fell earlier. Some are sitting, with their legs tucked under their bodies. They are chewing the cud, waiting for the snowfall to stop. Then they will set off to search for patches of grass and shrubs to eat. Their double humps are plump, a sign they are still feeding well.

There are six species in the camel family. Wild camels still live in the Mongolian steppes. These are the two-humped Bactrian camels. Most Bactrian camels, though, are domestic animals, used to carry goods or people. Arabian (one-humped) camels are also used by people. No truly wild ones remain. They were tamed thousands of years ago. They are used by people all over northern Africa and the Middle East.

The other members of the camel family are all South American. Two are wild. The smallest is the graceful vicuña, which lives high in the Andes. The larger guanaco lives lower on the mountains, and also down to sea level on some grasslands. The other two species are the llama and alpaca. They are both domestic animals. It is believed that people bred them from the wild guanaco starting 5,000 or more years ago.

DESERT SURVIVORS

Camels are able to survive in some of the harshest deserts. They wander widely, and feed on a variety of plants. They eat thorn bushes, dry vegetation, and even saltbushes. Most animals will not eat saltbush, but camels seem

to thrive better with some salty food. Camels can withstand long periods without water. They often graze far from oases, and have been known to go as long as 10 months without water. In such cases, a camel loses much weight and strength. Once it finds water, a thirsty camel may drink as much as 30 gallons within a short time with no ill effects.

Camels do not store water in their bodies. They are just good at keeping what they have got. They produce little urine, and dry droppings. They hardly sweat. Instead, they allow their body temperature to rise by as much as 17°F on a hot day, and cool down at night. This rise and fall in temperature would make most mammals ill, but not camels. Camels can sweat if they really need to. The inside of their large nose helps trap moisture rather than letting it escape from the body in the breath. Any moisture dripping from the nose runs down a groove to the split upper lip. The nostrils can be closed to keep out

desert dust. The fur in the ear-flaps also keeps out dust.

KEEPING COOL

The camel's fur coat acts as insulation from the heat of the Sun. The way the camel folds its legs right under its body when it is resting cuts down the amount of surface exposed to the Sun. The camel's humps are filled with fat. This is mainly a reserve of food, as is fat in most animals, but it is concentrated on the back. Here it also serves as a barrier to the Sun's rays, without wrapping the whole camel in a layer of fat.

KEEPING WARM

Camels are not always in hot surroundings. Even in the Sahara the desert nights can be freezing. In Central Asia it can be bitterly cold all winter. The woolly coat of a camel protects it against cold. The Bactrian camel has a very thick winter coat. In spring the coat is shed in lumps, giving the animal a ragged look.

▲**Members of the camel family** The
Bactrian camel (**1**) and the Arabian
camel, or dromedary (**2**) have two-toed
feet with pads which spread the weight
on sand or snow. The llama (**3**) has been
an important beast of burden in South
America for nearly 5,000 years. The
alpaca (**4**) is raised for its wool. The
vicuña (**5**) is from the high Andes.

HIGH LIFE

The guanaco can also live in desert conditions, but the speciality of the South American members of the camel family is living high in the mountains. The vicuña lives in alpine grasslands at 12,000 to 15,800ft above sea level. The llama and the alpaca also live high in the Andes, as do some guanacos. These animals all have blood which is especially good at taking up oxygen in the thin air. They also have woolly coats that keep out the cold.

HERD LIFE

In camels and in the vicuña and guanaco, the herd usually consists of a single adult male with a harem of females. There may be 5 or 6 females in a vicuña herd, and up to 15 in a camels', plus their young. The male does not tolerate rivals, and young males are pushed out of the herd as they grow up. In the vicuña the male defends a particular area in which the herd feeds. He may stand on a mound, keeping watch on his group, and ready to give the alarm.

Camels nearly always produce a

▼With their long necks, Arabian camels feed from bushes or from the ground.

►An alert guanaco stands on the pampas. The guanaco can often get enough water from its food without drinking. In Argentina, it is hunted for its pelt.

single baby. The newborn is able to walk and go with the herd after only a few hours. It lacks the hard pads of skin that adult camels have on their knees on which they rest on the ground. It is adult at about 5 years old. Llama and guanaco babies grow up faster. They are able to run almost the moment they are born. They feed on milk for only 3 months. Some are able to breed at 1 year old.

SPITTING AND BLOWING

All members of the camel family chew the cud. They bring the stomach contents up for a second chewing. They can also use this mechanism as defence against any animals (including people) that annoy them. Their ears go back as they bring up part of the stomach contents and spit the foul-smelling liquid over the enemy.

Male camels have another type of display which they use against rivals. In the mouth is a piece of skin which the camel can fill with air and blow out

►Even in thin mountain air, the vicuña can run at 28mph.

like a pink balloon, at the same time giving a "roar".

VITAL AND VALUABLE

The camel family has been vital to people. All species are marvellous beasts of burden in difficult conditions. A camel can carry 220lb of luggage 18 miles in a day. A llama can carry a 130lb burden nearly as far, high

In the early morning snow, Bolivian llama drivers prepare their animals for a day of high-altitude transport. The llama may be brown, black, white or blotched. As well as carrying goods and giving wool, llamas provide meat and leather, and their dried dung is used as fuel.

in the mountains. As well as working, camels provide wool and meat. They are also milked, and can give 10 pints a day for up to 18 months. They can be ridden, raced, used in warfare, and also used as wealth.

The llama was the mainstay of the civilization of the ancient Incas. When the Spanish reached South America in the 15th century AD, over 300,000 llamas were being used in the silver mines alone. Thousands of others carried goods. The llama has good wool, but the smaller alpaca is bred especially for its wool. Some animals grow wool almost to the ground. The llama is becoming less important as a pack animal, but the alpaca is still important for wool production. There are about 3,000,000 alpacas in Peru.

The finest wool of all comes from the vicuña. However, instead of trying to conserve this valuable animal, and harvest the wool, people have killed it for a single fleece. Numbers dropped from several million in the 1500s to less than 15,000 in the late 1960s. But now the vicuña is fully protected and the population is growing again. Now there are over 80,000.

MUSK DEER

A tiny spotted deer huddles in a mossy hollow beside a rock. It is a baby Musk deer. Its mother has left it for the day while she feeds in the forest. Now it is evening. The Sun still shines on the mountain peaks, but in the forests on the lower slopes it is becoming dark. Slowly and carefully, the mother deer returns to feed her baby.

Musk deer live over a wide area of Asia, from Nepal to Siberia. In spite of this, their life-style and habits are not well known. They are shy animals, and live in dense forests and brush where they are hard to see. Until quite recently only one species was recognized, but now there are thought to be three. Although called "deer", scientists have placed them in a family of their own as neither males nor females have antlers.

Male Musk deer, however, have a pair of large upper canine teeth that project downward from the lips by as much as 3in. These saber-like teeth can inflict deep, and sometimes deadly, wounds on the neck or back of a rival male. The females have much smaller canines that do not show when the mouth is closed.

MOUNTAIN DWELLERS

Musk deer live in mountainous regions and many live 9,000ft or more above sea level. Here it is often cold, and the deer have a coat of thick bristly hairs which are good at trapping heat. The back legs are about 2in longer than the front legs and are very muscular. Musk deer are good at jumping, and if they are disturbed they dart off with enormous bounds. They stand on two toes of each foot, but the other two toes are well developed and sometimes help to give grip. Musk deer can run well on snow or mud. They can also climb well on crags and cliffs, rocks, and even trees.

Apart from mothers with young, Musk deer are usually solitary. They stay in a particular area, and use the same trails and feeding places regularly. They also deposit their droppings in specific places, and scratch earth over them with their front legs. Males produce an oily secretion from a tail gland. They use this to mark tree trunks, twigs and stones within their area, to show ownership.

During the mating season, which is usually in midwinter, the male Musk deer run about restlessly and chase after the females to mate with them. The males fight one another for the females. After the mating season the males return to their original areas.

PRICELESS POUCH

For thousands of years, Musk deer have been hunted and killed for musk, the substance that gives them their name. Musk is a jelly-like oily secretion with a strong smell. It dries

▼The long protruding canines of the male Musk deer give it a mournful expression.

MUSK DEER Moschidae
(*3 species*)

Habitat: dense forest, coniferous or deciduous.

Diet: leaves, flowers, young shoots and grasses, twigs, mosses and lichens.

Breeding: 1 young after pregnancy of 150-180 days.

Size: head-body 2½-3½ft, plus 2in tail; weight 15-37lb.

Color: grayish-brown to golden, speckled.

Lifespan: 13 years.

Species mentioned in text:
Musk deer (*Moschus moschiferus*)

▲Musk deer have large mobile ear-flaps and good hearing. In the thick vegetation of their home hearing is probably more useful than eyesight.

▶A newborn Musk deer is tiny compared with its mother. For the first weeks of its life it stays in a hiding place.

to a black powder. It is produced by mature males. The animals develop a large gland, a pouch about the size of a clenched fist, on the lower belly. In this, musk is produced.

Musk is used in Europe as a base for perfumes. In the Far East it is used in many kinds of folk-medicines. It is much in demand, and at times has been literally worth more than its weight in gold. But it is not necessary to kill the males to obtain musk. It can be scooped from the pouch when the animals are sexually most active. In China people have begun keeping and breeding Musk deer so they can be farmed for the musk.

CHEVROTAINS

A long tunnel winds through the vegetation on the jungle floor. A small animal runs along its length. It is a chevrotain. It pauses as the tunnel reaches a jungle clearing and peers out cautiously. It can detect no enemies, and emerges into the clearing. It sniffs and tastes some fruit fallen from the trees above. It eats a few of the best. Then, as quickly as it came, it darts into another tunnel and disappears.

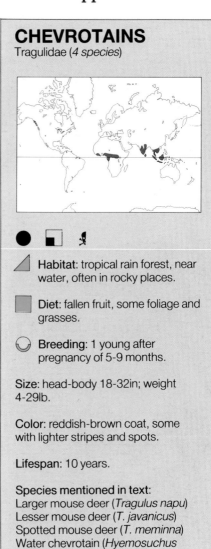

CHEVROTAINS
Tragulidae (4 species)

● ■ 🗡

◢ Habitat: tropical rain forest, near water, often in rocky places.

■ Diet: fallen fruit, some foliage and grasses.

◖ Breeding: 1 young after pregnancy of 5-9 months.

Size: head-body 18-32in; weight 4-29lb.

Color: reddish-brown coat, some with lighter stripes and spots.

Lifespan: 10 years.

Species mentioned in text:
Larger mouse deer (*Tragulus napu*)
Lesser mouse deer (*T. javanicus*)
Spotted mouse deer (*T. meminna*)
Water chevrotain (*Hyemosuchus aquaticus*)

There are four species of chevrotain. One lives in the tropical forests of equatorial Africa. The three others live in the tropical forests of India and South-east Asia. Chevrotains are small animals, the smallest of hoofed mammals. The tiniest, the Lesser mouse deer, is no larger than a common rabbit. The alternative name for chevrotains is "mouse deer," but these are not true deer.

MIDGET MALES
Chevrotains are rather unusual among mammals in that the females are generally larger than the males. In the Water chevrotain, females are 20 per cent larger. Chevrotains have tubby bodies, but each of their short legs may be no thicker than a pencil.

Neither sex has antlers, but both have narrow, curved and sharp upper canine teeth. In males these teeth stick out from the lips. In the breeding season the animals sometimes use them in fighting, to bite at the body of another male.

CLOSE TO WATER
All chevrotains live near to water. The African species, the Water chevrotain, is a good swimmer. It can dodge enemies by jumping in the water, submerging, and coming out of the water some distance away under overhanging branches. Many kinds of

▲ The Spotted mouse deer lives in India and Sri Lanka. It is active mainly at night, running along jungle trails.

▶ The Lesser mouse deer uses its large eyes and sensitive nose to be alert to enemies approaching.

animals, from snakes to jungle cats, will eat chevrotains.

IGNORING STRANGERS
Chevrotains are very solitary animals. Except in the breeding season, they show little interest in one another. Each animal stays in one area, its home range. Other chevrotains of the same sex keep out of it, but there is little evidence of fighting off intruders. The Water chevrotain marks its home range with urine and droppings. The animal has special anal glands that add scents to the droppings. Home ranges can be up to 70 acres in males, but only half that for females. Where food is plentiful there are as many as 28 chevrotains in a square mile. Although common animals, they are shy and rarely seen.

The Larger mouse deer and the Water chevrotain breed throughout the year. A single baby is born, and left in a hiding place. The mother returns only to feed it on milk. Chevrotains grow up fast. By 5 months of age a Larger mouse deer is fully grown and sexually mature.

DEER

It is fall. In a woodland glade a White-tailed deer buck is surrounded by does. He lifts his head and roars. From beyond a ridge a rival replies, and then comes trotting over the hill. The two bucks walk in the same direction for a while, gauging each other's size. Then they turn towards one another and their antlers clash. They push back and forth in a trial of strength. The first buck is just the stronger. The other buck breaks away and runs, but is chased by the victor.

Deer live in North and South America, Europe, North Africa, and over most of Asia including many islands. There are no deer in Africa south of the Sahara Desert. Nor are there deer in Australia and New Zealand except for those introduced by people.

Deer range in size from smaller than an Alsatian dog to the moose which towers over an adult human, but most are medium-sized animals. The majority of deer are colored in a shade of brown. Some kinds are spotted. The dull colors are good camouflage for animals such as deer, which are good to eat and often hunted by others. Deer mostly feed on grasses and low-growing plants, but some kinds feed on leaves and twigs from bushes and trees. They all chew the cud.

WHAT MAKES A DEER?

Deer look similar to other plant-eaters, especially antelopes, and have graceful, elongated bodies, slender legs and necks, and short tails. They have long heads with jaws that bear many chewing teeth to deal with plant food, and also large noses providing a keen sense of smell. Their sense of hearing is also good, and the large ear-flaps are very mobile. The large round eyes are set on the side of the head, giving a good all-round view to warn of approaching enemies.

The feature which sets deer apart from all other animals, however, is the possession of a pair of antlers. These are carried only by the males (except in caribou). Antlers are made entirely of bone, unlike the horns of cattle or antelope, which have a horny covering over a bony core. In some deer the antlers are just simple spikes. In many they are branched. The way the antlers

▶ A Red deer stag roars to lay claim to his harem. The thick neck and mane develop for the breeding season.

▼▶ American species of deer The Southern pudu (*Pudu pudu*) (1) is the smallest deer, and comes from forests in South America, as do the Red brocket (5) and the huemul (*Hippocamelus antisensis*) (4). The Swamp deer (3) is the largest South American deer and lives on wet grasslands. The Pampas deer (*Ozotoceros bezoarticus*) (2) lives on dry plains. The White-tailed deer (6) lives in small herds from North to South America.

DEER Cervidae (*36 species*)

Lifespan: 10-20 years.

Species mentioned in text:
Chital (*Axis axis*)
Fallow deer (*Dama dama*)
Mule or Black-tailed deer (*Odocoileus hemionus*)
Père David's deer (*Elaphurus davidiensis*)
Red brocket (*Mazama americana*)
Red deer (*Cervus elaphus*)
Reeve's muntjac (*Muntiacus reevesi*)
Roe deer (*Capreolus capreolus*)
Rusa deer (*Cervus timorensis*)
Sambar (*C. unicolor*)
Sika deer (*C. nippon*)
Swamp deer (*C. duvauceli*)
Tufted deer (*Elaphodus cephalophus*)
Wapiti or elk (*Cervus canadensis*)
Water deer (*Hydropotes inermis*)
White-tailed deer (*Odocoileus virginianus*)

◖ ■ ⚔

△ Habitat: mostly woodland and forest; some found on tundra or open grassland.

▢ Diet: grasses, or shoots, twigs, leaves and fruit of shrubs and trees.

◎ Breeding: 1 or 2 young after pregnancy of 24-40 weeks.

Size: smallest (pudu): head-body 2½ft, shoulder height 1¼ft, weight 17lb; largest (sambar): head-body 6½ft, shoulder height 4½ft, weight 600lb (but see Moose).

Color: mostly shades of gray, brown, red and yellow. Some adults and many young spotted.

branch is slightly different for each species, so is a good means of identification. In a few species, such as the Fallow deer, the antlers have flattened sections with a hand-like (palmate) appearance.

The main function of antlers seems to be for fighting and display between the males. As males mature the antlers become bigger and, in some cases, more complicated in branching. They show well the maturity, strength and condition of a male. Thus they are good signals which another male can interpret before deciding on a challenge for supremacy.

RUTTING

In the breeding or rutting season, in species such as the elk, males (bucks or stags) fight for a group or harem of females (does or hinds). Fights mostly take place between well-matched males. Smaller animals give up their challenge on hearing the herd leader's voice or seeing his size. Fighting mainly consists of pushing matches with antlers locked, which may go on for many minutes. Losers are usually allowed to run away without harm, but sometimes bad injuries are caused by antlers. Fighting and guarding a harem takes up much energy, and males become very tired by the end of the rutting season. They need a period of rest and building up before they are again ready to mate.

RENEWABLE WEAPONS

During this rest period the antlers are shed. All deer replace the antlers regularly, unlike the permanent horns of antelope. The bone of each antler dissolves away at the base, and the antlers drop off, usually within a day or two of each other. A bony stub is left on each side of the skull. The deer begins the process of growing a new set. This involves making large amounts of new bone. Many deer gnaw cast antlers, so getting back some of the minerals they need.

While the antlers are growing they are covered with skin and hair. This covering is known as "velvet". So much chemical activity goes on below the velvet that it can feel hot to the touch. When the antlers have grown to their full size for the year, the blood supply to the velvet is cut off and the skin dies and begins to shrivel. The deer may help get rid of the dead skin by rubbing its antlers against a tree.

HERDS

For much of the year typical deer live in single-sex herds. A group of females and young keep together. Males may be found singly, or else in a "bachelor herd." The intense rivalry between males occurs only in the breeding season.

The size of herds varies according to species and the habitat they live in. The Red deer is basically a woodland

▲Fallow deer prefer open woodlands. Originally from the Mediterranean, they have been introduced to Britain, and more recently to Australia.

◄Two male Black-tailed deer struggle for supremacy. Their branched antlers lock together.

►Out of the breeding season, Red deer stags live together in herds without fighting. These stags are "in velvet," or in the process of growing new antlers.

animal, and there it lives in herds of about 20. In open country the herds are often larger, up to 100 strong. The chital, a deer that feeds on grass and lives in grassland and light woodland in India, may have herds of more than 100. At the other extreme is the Roe deer of Europe and Asia, and the Red brocket of South America, each of which often inhabits dense woods and forests. These deer mainly browse, eating shoots, twigs and leaves. They are solitary animals, except when breeding.

SIGNALS

Many species of deer have light-colored rumps or tails. When the animals flee, these marks are easy for

▲ The elk, or wapiti, is found in western North America and Asia. It is closely related to the Red deer, but larger. Its antlers can be more than 3ft long.

others in a herd to follow. Other signals are also used. Mothers and young sometimes bleat to keep in touch with one another. Many deer "bark" when they are alarmed. Males in the breeding season can be noisy too. Their sounds range from a bellowing roar in the Red deer to a whistling scream in the Sika deer.

Deer have sensitive noses, so scent signals are especially important to these animals. Many species have special glands between the toes, which leave behind their owner's scent imprint. Glands in slits just in front of the eyes produce strong-smelling secretions. These the animals deposit on twigs and grass stems.

REARING THE YOUNG
Most deer produce a single young (calf). This stays hidden for a few weeks, emerging only when the mother visits to nurse it. After this period it moves with its mother and herd. It nurses for several months. A young deer may still be following its mother while she is rearing her next calf. Babies often have a spotted coat, but this pattern disappears long before adulthood. A few kinds of deer, including the Fallow deer, chital, and Sika deer, are spotted as adults.

The Water deer of China is unusual in that it often gives birth to triplets. Even bigger litters have been recorded. This solitary swampland deer is also unusual in having no antlers in either sex. The males have tusk-like canine teeth, as do the males of the muntjac and Tufted deer, both of which have very small antlers.

DEER AND PEOPLE
Deer have been hunted by people for their flesh (venison), skins and antlers since prehistoric times. Deer are still hunted, but in some places farms have been started to produce venison. In parts of Asia deer damage crops and are considered a pest.

Some 70 years ago, Père David's deer was nearly exterminated. This is a large species with spreading hoofs, native to the river plains of northern China. When Europeans first visited China in the 13th century, the only specimens left were in the Emperor's hunting park in Beijing. The last one there died in 1920. The species was saved because a few had been brought to Woburn in England. These bred, and now some have been taken back to China.

▲The Rusa deer of Indonesia shows the large ears, big eyes and wet nose typical of deer.

▼European and Asian species of deer The Roe deer (1) is found over much of Europe and Asia. Spotted deer include the chital (2) and Sika deer (4) from Asia. Reeve's muntjac (3), and the Tufted (5), Père David's (6) and Water deer (7) are all from Asia.

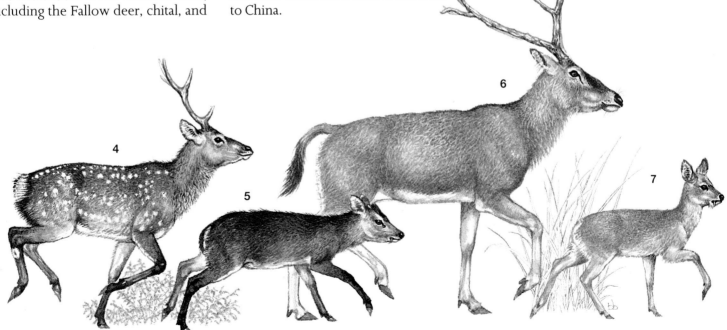

MOOSE

The bull moose flounders through a deep snowdrift on his long legs. A pack of wolves closes in behind him. He reaches solid ground, but he is too tired to run. He turns and faces his pursuers. As the first wolf comes within range he kicks out with one of his front feet and sends the attacker flying. He hits three more with flailing hoofs. The pack turns and gives up the hunt.

The moose is the largest deer. It is found in Canada, Alaska and parts of the north-western USA. It also lives from Scandinavia across northern Europe and Asia to northern China. It varies a little in size and color from place to place, but there is just a single species.

In Europe this species is called the elk. In North America it is called moose. Unfortunately, another deer, the wapiti, is often called an elk in America. So talking about "elk" can be confusing, unless somebody makes clear what they mean. Here "moose" has been used for American, European or Asian animals.

GIANTS OF THE SWAMPS

The moose lives in wooded areas, but particularly likes moist conditions where woods are sprinkled with lakes, ponds and swamps. The moose has broad hoofs which spread its weight on soft earth or mud. With its long legs it can reach high into bushes or tree branches to browse on leaves and shoots. Standing on all four legs, a big moose can reach up 10ft high. Sometimes it may rear on its hind legs to pluck food from 13ft up. A moose will also use the weight of its head or body to bend a sapling so its shoots are within reach. The moose has long and mobile lips to pluck its food. It will eat the equivalent of 20,000 leaves a day. It feeds on a variety of trees, including ash, birch and willow.

In the warmer months of the year a moose goes into pools to feed on water plants. These are rich in the minerals it needs as well as in energy. In a summer day a moose may eat 1,100 water plants. These may include water lilies, pondweed, horsetails, bladderworts and bur-reed. A wading moose often submerges its whole head as it feeds. Sometimes the whole animal goes underwater.

The moose is not built for eating grass, but will occasionally kneel on its front legs, with its chin sweeping the ground, and crop grass. In winter, when there is little else to eat, it browses on conifers (cone-bearing trees) and eats their needle-like leaves.

The animal is usually solitary. Sometimes it is found in small groups. In spite of its great demand for plant food, an adult moose spends much of the year in a small area it knows well.

SPREADING ANTLERS

The moose, with its large size, drooping lips, and "bell" (the flap of skin hanging under the throat) is unmistakeable. The males have large antlers. These are palmate, with big flattened areas between the spikes. The record span is nearly 6½ft.

The moose is generally a quiet and cautious animal, but in the mating (rutting) season males are always ready to fight one another, and sometimes charge other animals or humans. Rutting is generally in September and October. The big antlers are grown and ready for use by August. By the time winter arrives, the bulls (males) no longer fight. They may even form little groups together. Their antlers fall, and new ones do not grow until spring.

Winter is the most dangerous time for moose. Snow slows them down, and neither sex has antlers to fight off predators. But they defend themselves with their hoofs. Snow also cuts down the amount of food available.

TWIN BIRTHS

Pregnant moose give birth during the summer. Often there are two young born. Unlike many deer, the babies do not have spotted coats. They stay with the mother for at least a year, sometimes two.

►A female moose (cow) and her calf feeding on water plants in summer. In winter moose feed mainly on land plants.

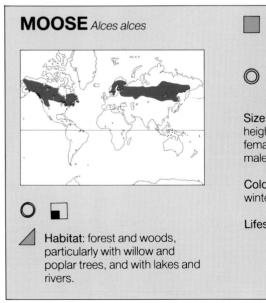

MOOSE *Alces alces*

■ Diet: leaves of shrubs and trees, bark and twigs, aquatic plants, some ground-living plants.

◎ Breeding: 1-3, usually 2, calves after pregnancy of 240-250 days.

Size: head-body 8¼-10ft; shoulder height to 7½ft; weight 880-1,760lb, females about 25 per cent smaller than males.

Color: brown, darker in summer than winter. Young unspotted.

Lifespan: 20 years.

○ ◻ ◣ Habitat: forest and woods, particularly with willow and poplar trees, and with lakes and rivers.

CARIBOU

More than 1,000 deer trek across the flat landscape. It is spring. The animals are migrating north. Much of the time the whole herd moves at a steady trot. They come to a river, broad and flowing fast. They swim across and land downstream on the other side. On they go to the summer feeding grounds.

CARIBOU
Rangifer tarandus

○ ■

◢ Habitat: Arctic tundra, northern woodland edges.

◼ Diet: lichens, dwarf shrubs, grasses, sedges.

◎ Breeding: 1 young after pregnancy of 210-240 days.

Size: head-body 4-7½ft; shoulder height to 4½ft; weight 200-600lb; males larger than females.

Color: brown in summer, gray in winter, with white rump. White mane in rutting males.

Lifespan: 15 years.

▶ This caribou bull has just shed the velvet covering its antlers. The brow prongs (tines), which come forward over the nose, can be useful in winter to brush snow off a patch of food.

Caribou live around the edge of the Arctic, on the northern continents, and on islands such as Spitzbergen and Greenland. The same species is found all over this area, with some variation in size and color. In Europe this species is known as the reindeer. In northern Norway, Sweden and Finland, Lapps tend herds of partly tame caribou.

ALL ANTLERED
The caribou is the only deer in which both sexes have antlers, often with palmate tops. Those of males are much larger and have more points. Females do not use their antlers to fight one another, but the antlers are useful to them in the winter for scraping through the snow to find food beneath. Caribou also use their big feet to dig for food. The name caribou means "shovel-foot" in a native American language.

MIGHTY MIGRATIONS
During the winter caribou live at the northern edges of the woodlands bordering the Arctic. In spring they migrate far into the tree-less tundra. Here they feed on lichens and on grasses, sedges and low-growing shrubs during the brief Arctic summer when these plants grow fast. In the fall the caribou migrate back again to the shelter of the trees for the winter. Here they scratch a living from caribou "moss" (a kind of lichen), and buds and shoots of shrubs.

On migration caribou often travel in herds several thousand strong, and cover 100 miles in a single day. They follow definite "trails" that have been used by the herds for many years.

▼Caribou hoofs are broad and flat. They are good for walking on snow or soft ground. When caribou walk their feet make a clicking sound, caused by a tendon moving over an ankle bone.

◄Caribou grazing in the tundra. Here the animals make use of the short summer season of plant growth. In 1986, as a result of the Chernobyl nuclear disaster in the USSR, tundra plants became poisonous to caribou. Many animals died or became unfit for food.

Migration routes can be 660 miles long. Wolves and other predators follow the deer on their migrations.

Not all caribou migrate. In some areas they stay in woodlands all year.

FURRY NOSES

In almost all deer the end of the nose is smooth and wet, but in the caribou it is fur-covered. This is an adaptation to keep in body heat in Arctic conditions. The caribou's coat has a very thick woolly underfur, guarded by longer straight hollow hairs.

Young caribou are born during the summer. The good feeding on the summer pastures lets them build up strength for the fall migration and cold of winter. It also prepares the adults for breeding. Rutting takes place in September and October. Males (bulls) may fight fiercely and gather a harem of up to 40 females (cows). After the rut the antlers drop off.

PEOPLE AND CARIBOU

Early in this century millions of caribou each year migrated back and forth across the tundra of North America. Now the animals are numbered in thousands rather than millions. Forest fires caused by people in the caribou's winter areas seem to be one reason for their decline.

Caribou in Europe have been tamed for hundreds of years. The Lapps follow the herds, using them for most of their needs, including milk, meat and clothing. They mark the animals, and castrate (remove the reproductive organs of) males not needed for breeding, so cutting down fighting. But the animals roam the tundra pastures as if still wild.

GIRAFFE AND OKAPI

Two female giraffes feed on a thorn tree. A mile away a lone male crops another tree. He is disturbed. He stops feeding and watches a movement in the distance. The females stop feeding and watch him. Uneasy, the male moves away from his tree. The females, and then three more giraffes, follow him to a safer place.

GIRAFFE AND OKAPI
Giraffidae (*2 species*)

○ ■

◑ **Habitat**: open woodland and savannah (giraffe); dense forest (okapi).

▢ **Diet**: leaves, bark and shoots. Some flowers, seeds and fruits.

◎ **Breeding**: 1 calf after pregnancy of 453-464 days.

Size: (giraffe) head-body 12½-15½ft, height to 18ft, weight 1,210-4,240lb, males larger than females; (okapi) head-body 6½ft, height to 5½ft, weight 460-550lb.

Color: (giraffe) red-brown to almost black patches of variable size separated by network of lighter fur; (okapi) dark velvety purplish-brown, with white stripes on rump and legs.

Lifespan: 15-25 years.

Species mentioned in text:
Giraffe (*Giraffa camelopardalis*)
Okapi (*Okapia johnstoni*)

The giraffe family consists of two species, the giraffe and the okapi. The giraffe lives in most of Africa south of the Sahara. The okapi is found in a small area of rain forest in Zaire, but is locally common. Although a large animal, it was unknown to Europeans until 1901. Little is known of its behavior in the wild.

The giraffe is the tallest animal. The biggest ever measured was 20ft high to the top of the horns. The males (bulls) are heavier and much taller than the females (cows), which rarely exceed 15ft in height.

CHOOSING DINNER

The two sexes also have different styles in eating. Male giraffes stretch up high into a tree for their food. Females more regularly feed at around their shoulder height. So not only does the giraffe feed above the level of most other animals, but the food available is divided between the sexes.

The giraffe is a browser, picking leaves and shoots off trees and shrubs. It may also eat seed pods, flowers, fruits and climbing plants. The giraffe has very mobile lips which it uses to

▲Young bull giraffes hold ritual fights to discover which is stronger. They use their necks for wrestling, and their horns and heads as butts. Only the strongest adult bulls mate with the females in an area.

▼The okapi is a secretive, solitary animal. It feeds on leaves of young tree shoots. Its face and tongue are like the giraffe's. It has poor sight, but good senses of hearing and smell.

pull food to its mouth. It can stretch out its tongue about 18in to gather food. The canine teeth are shaped like a comb, and are used to strip leaves from a branch. The animal often feeds from thorn trees, which have spines several inches long. It can pick leaves from between these, but will even chew thorns if they are taken into the mouth.

LONG NECK AND GIANT HOOFS

Many features of the giraffe's anatomy are very odd. Each giraffe has its own individual coat pattern, like human finger-prints. There are, though, different types of color and pattern according to the part of Africa the giraffe lives in. The enormously long neck has only seven bones in it, like the neck of other mammals, but each

one is greatly elongated. The body is comparatively short in length. Tangling the long legs is avoided by moving both legs on the same side of the body together. This produces a loping movement. When the animal raises or lowers its head, blood drains or rushes to the brain. To cope with this the blood vessels are specially elastic to help pump blood, and strong valves in the veins prevent backflow.

A giraffe is born with small horns. As it gets older, particularly if it is a bull, these grow thicker and heavier, and many bony lumps appear on the skull as well. The giraffe also has soup plate-sized hoofs. These it uses as weapons; with a powerful kick, a giraffe can kill a lion. The skin is thick and tough for defence. Adults have few enemies, but about half the young (calves) die in their first year, killed by lions, leopards or hyenas.

The giraffe has good hearing and sense of smell, but it closes its nostrils when poking its head into a thorn tree. Its most acute sense, though, is sight. It can see clearly for many miles across the plains.

CALVING

A pregnant giraffe gives birth to her baby in a special calving area within her home range. The baby can stand soon after birth. After a week or two the calf may join up with other calves and form a "nursery group." The group is left alone during the middle of the day while the mothers go off to feed. A newborn giraffe is 6ft tall. In its first year, it may grow 3in a month. Giraffes are adult at about 5 years old, but males may be 8 years old before they manage to breed.

◄This male reticulated giraffe, feeding at full stretch, belongs to a race from northern Kenya and Somalia. To pump blood to its head it has a heart 2ft long and weighing more than 20lb.

WILD CATTLE

WILD CATTLE Bovidae, Tribe Bovini (*12 species*)

○ ■ ♒

◢ **Habitat:** forest, glades, savannah, prairie.

■ **Diet:** grasses, other plants.

◎ **Breeding:** 1 calf after pregnancy of 254-340 days.

Size: smallest (Mountain anoa): head-body 5ft, shoulder height 2¼ft, weight 330lb; largest (gaur): head-body 10ft, height 6½ft, weight to 2,000lb.

Color: black, dark or reddish-brown, some with white markings.

Lifespan: 20 years.

Species mentioned in text:
African buffalo (*Syncerus caffer*)
American bison (*Bison bison*)
Banteng (*Bos javanicus*)
Domestic cattle (*B. primigenius*)
European bison (*Bison bonasus*)
Gaur (*Bos gaurus*)
Kouprey (*B. sauveli*)
Lowland anoa (*Bubalus depressicornis*)
Mountain anoa (*B. quarlesi*)
Tamarau (*B. mindorensis*)
Water buffalo (*B. arnee*)
Yak (*Bos grunniens*)

A huge shaggy brown beast emerges from a forest in eastern Europe. It is a male bison. He snorts, and clouds of water vapor come from his nostrils. He looks for a dry open patch of ground. Suddenly he lies down, and with great energy twists and rolls from side to side in the dust. He gets to his feet again and shakes himself. Then, with nostrils still steaming, he walks back in among the trees.

There are 11 species of wild cattle. Most live in Asia, Africa and Europe. Just one species, the last of the tribe, the bison (the "buffalo" of the Wild West), is found in America.

FOREST FORMS

Because domestic cattle are kept in fields of grass, it is tempting to think that wild cattle live in grasslands. In fact, most kinds of wild cattle also live in forests or woodlands, feeding on grasses in leafy glades or at forest edges. The European bison is a forest dweller that feeds on leaves, twigs and bark of trees.

The American bison lives on prairies, but could once be found in forests too. It feeds mainly on grass.

▲Yaks find enough food to live even in the snows of Tibetan mountains. Like other cattle, their food ferments in their stomachs. Heat from this (about 103°F) provides "central heating" for the body.

◀American bison graze on grassland. In North America, European settlers slaughtered bison for meat or to deprive the local Indians of their sole means of support. Today many of the animals are protected in reserves.

The African buffalo grazes on soft grasses. There is a large dark-colored form which is found in savannahs and open woodlands, and a smaller, redder form which is found in tropical forest.

The yak is unusual in living in grassland, tundra and ice deserts at heights of over 13,000ft in Tibet. It has a dense undercoat and long shaggy dark hair to help keep out the cold.

FEELING FOOD

Cattle have long jaws and good chewing teeth. The end of the snout is wide, with a bare pad of skin around the lips, and large nostrils. Wild cattle have a keen sense of smell that they use to detect enemies and find food. While they are grazing, they constantly sniff the pasture to find the best fodder. Bare wet lips also provide a good sense of touch and taste, and help them find the most succulent parts of the plants. Their senses of sight and hearing are good, but not sharp.

When alarmed, cattle give an explosive snort, quickly followed by an alarm posture in which the head is held high. They face the danger with the body tensed. In the African buffalo, an alarm call may bring the whole herd to help a frightened animal.

HORNS

All the cattle are well equipped with horns. They are best developed in males, which use them to fight one another, but they are also good weapons for defence. In the African buffalo the lumps of bone to which the horns are attached make a massive ridge across the skull, and the horns meet in the mid-line, giving a helmeted appearance. A solitary buffalo may be attacked, but in a herd these animals are safe. Blind, lame or three-legged buffalo have been known to survive within the herd, and even lions are not safe from the herd's charge.

The longest horns of all are found in the Water buffalo of Asia. A specimen shot in 1955 had horns which measured 14ft from tip to tip along the curve. Anoas and tamaraus are small species of buffalo that live in thick forest on some islands of Indonesia. They have straight, backward-pointing horns that allow them to move easily through the undergrowth.

HERDS

Little is known of the social life of the smaller buffalo, but in all other

47

species of cattle group- or herd-living is well developed. Females spend their lives in a group with other cows and their calves. Young bulls generally leave the group when about 3 years old. Forest forms such as the gaur, banteng and African forest buffalo live in groups of up to 10. Bison groups can be up to 20 strong. Groups may join up during the breeding season to form large herds. A mature bull may stay with a group through the year, or bulls may live alone or in male groups and join the cows just for mating.

In those species that live in open country, several cow groups may stay together all year. Their herds can number as many as 350 in African buffalo living on the savannah. In all cattle herds, the animals tend to do much the same thing at the same time.

FIGHTING FOR FEMALES

In the breeding season mature males keep close to females that are ready to breed, and try to keep other males away. Often they do this by threats. In American bison threats include bellowing and rolling in the dust. Approaching an opponent head-on,

◄An African buffalo displays its huge horns. When hot, it goes into shade or wallows in mud to keep cool. The oxpeckers riding on its back pick insects and other pests from its skin.

▼Two bison bulls clash in a battle for mastery of a herd. Each can weigh 1,650lb and has an enormously strong head and shoulders.

or standing tall sideways on, may also warn him off.

If all else fails, the bulls fight. They slam their heads together, their hoofs churning up clouds of dust. Clumps of hair bigger than a person's fist are knocked from their heads by horns grinding against one another. The animals circle, trying to dodge lunges and drive a horn into the opponent's ribs and flank. In such a fight, one contestant may die of his wounds. But most disputes are settled peacefully.

TAME TYPES

More than one species of wild cattle has been tamed by people. The aurochs, which died out as a wild animal in 1627, was bred to produce the domestic cattle of Europe and also the zebu of the tropics, a domestic type with a hump of fat on its shoulders. The Water buffalo, too, has been tamed for thousands of years, and worked in the paddy fields of Asia. There are now many more tame Water buffalo than truly wild ones. The yak has been turned into a valuable beast of burden in the Himalayan region, and also provides milk and butter. The largest of cattle, the Asian gaur, and the smaller banteng, have also been domesticated.

The usefulness of cattle to humans is one of the main reasons why most wild species are disappearing. Land which supports them is often taken over by people to use for domestic cattle. Most endangered is the kouprey, just a few of which are left in Indo-China.

PRONGHORN

A herd of pronghorns is grazing on the prairie. From time to time an animal lifts its head to watch for danger. One catches sight of a coyote slinking across the skyline. The pronghorn snorts in alarm, and raises the hairs on its rump, showing a bright white patch. At the same time scent is released from glands on the rump. Instantly the rest of the herd is alerted. Showing their rump patches, the whole herd rushes away from the hunting coyote at great speed.

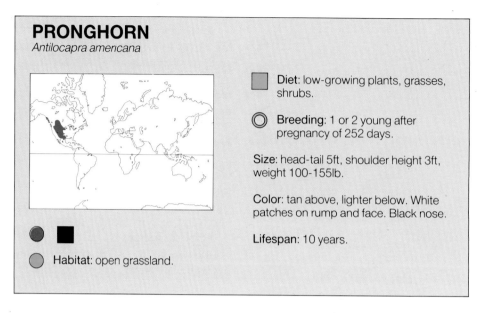

PRONGHORN
Antilocapra americana

■ Diet: low-growing plants, grasses, shrubs.

◎ Breeding: 1 or 2 young after pregnancy of 252 days.

Size: head-tail 5ft, shoulder height 3ft, weight 100-155lb.

Color: tan above, lighter below. White patches on rump and face. Black nose.

Lifespan: 10 years.

● ■

● Habitat: open grassland.

The pronghorn is related to both antelopes and deer. It is found in south-west Canada, northern Mexico, and the western USA. It is unusual in that both males and females have a pair of horns that are shed every year. Each horn has a bony core, and over this grows the outer layer, which is shed after the rut. The horn is branched, with a forward-pointing prong, from which the animal gets its name. The females have smaller horns than males, and may not show a prong.

SPEEDY SPRINTER
The pronghorn is the fastest runner among American animals. It can run at 50mph for nearly a mile. But it can also keep up a good speed for a long distance. It has been timed over 3 miles at a constant 35mph. Its chunky body bears long slim legs which have pointed hoofs cushioned to take the shock of strides reaching 25ft at full run. On the scrublands where it lives there is little cover, and speed to outrun enemies is vital.

SEEING AND SIGNALLING
Pronghorn eyes are unusually large and are set out from the skull, so that the animal can see all round. Long, black eyelashes act as sun-visors. The pronghorn can see well, and uses visible signals to keep in touch with others. On its brown body are some contrasting "flags." The white rump

▼ Pronghorn herds are small in summer, but may contain 100 or more animals in winter. They migrate with the seasons.

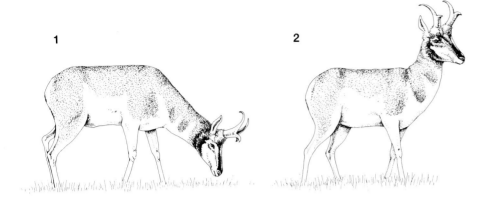

▲Pronghorns browse on vegetation at or just above ground level (1). They constantly keep watch for enemies (2), such as the coyote and the bobcat.

▼The pronghorn is built for running. It has a wide windpipe and large lungs, and its heart is twice as big as that of a sheep of similar size.

can signal alarm. The black horns and the black snout and throat are especially developed in males, and help distinguish the sexes.

The pronghorn has many scent glands, too, that can give information to other pronghorns. All have glands on the rump and between the toes. In addition, males have a gland above the tail and two beneath the ears. Scent from these is used to mark territories and during courtship to mark females.

RUTTING

The mating season is in the fall, but well before this each adult male stakes its claim to an area of the prairie. This may be as small as ⅓ sq mile, but usually contains some of the best food plants. At rutting time the male gathers on its territory a herd of up to 15 females and mates with all of them. An especially dominant male may be able to hold its territory for up to 4 years before being replaced, and sire a quarter of all fawns each year.

Fawns are born in early summer. Twins are common. A 2-day-old fawn can run as fast as a fully grown horse, but not for far. Usually, for the first 3 weeks, fawns hide, lying very flat to conceal themselves in the low vegetation. They are nursed for about 5 months. They mature quickly, and are ready to breed in the rutting season of the year after birth, although males are not usually strong enough to hold a territory until about 5 years old.

BACK FROM THE BRINK

When European settlers arrived in North America there were probably 50 million pronghorns living on the plains. Unregulated hunting and disturbance reduced the numbers to 13,000 by 1920. The pronghorn has a keen sense of curiosity, and people who wished to kill them lured them by waving a handkerchief. The animal was on the verge of extinction. Efforts to conserve it, though, have succeeded, and now there are 450,000.

ELAND

A herd of eland lies in the shade of some bushes in the heat of the day. As the Sun's heat lessens, they get up and begin feeding. They eat the tastiest leaves, then set off to find another patch of shrubs.

The eland is one of a group of nine species of large African antelopes with spiral horns. They have some similarities to the cattle, to which they are related, but are more slender, with narrower faces. The horns corkscrew back from the top of the head.

GIANT ANTELOPES

There are two species of eland. The Common eland lives in grassland and open woodland in East, Central and southern Africa. Both sexes have horns, but they are bigger in males. The eland has a massive cow-like body, but is agile enough to jump high fences. It lives in herds that are mostly small, but in the breeding season can have 100 or more animals in them.

▼ The Four-horned antelope is a small Indian species. Only males have horns. It is a distant relation of eland and gave rise to modern cattle.

ELAND Bovidae, Tribe Strepsicorni (*9 species*); also Tribe Boselaphini (*2 species*)

Habitat: forest, open woodland and woodland edges, grassland.

Diet: leaves, fruits, flowers, seed pods, bark, tubers, grass.

Breeding: usually 1 young after pregnancy of 245-270 days.

Size: smallest (4-horned antelope): head-body 3½ft, shoulder height 2ft, weight 44lb; largest (eland): head-body 11ft, shoulder height 6ft, weight 2,040lb.

Color: fawn, gray, or reddish-brown; striped in forest forms.

Lifespan: 15-20 years.

Species mentioned in text:
Bushbuck (*Tragelaphus scriptus*)
Common eland (*Taurotragus oryx*)
Four-horned antelope (*Tetracerus quadricornis*)
Giant eland (*Taurotragus derbianus*)
Greater kudu (*Tragelaphus strepsiceros*)
Mountain nyala (*T. buxtoni*)
Nyala (*T. angasi*)
Sitatunga (*T. spekei*)

▲ Male elands "horn-tangle" to see which is strongest and will mate with the females. They thresh shrubs (1) and dig horns in mud (2), coating them with smelly mud and sap prior to fighting (3).

The herds are often found alongside giraffes or zebras. Elands wander, rather than staying in a fixed place.

The Giant eland, in spite of its name, does not differ much in size from the Common eland, but it is more a woodland species, with populations across northern Central Africa.

The other members of this group include the Greater kudu. Although males may be 4ft at the shoulders, with horns 6ft long, they are very graceful animals, with a bluish gray-brown coat with white vertical stripes. Females are smaller and do not have horns. The kudu likes woodland, especially in hilly regions of eastern and southern Africa.

The nyala lives in south-east Africa in dense thickets and riverside vegetation. The Mountain nyala lives on heathland and in mountain forests in some areas of Ethiopia. Once thought to be rare, it is now known to be reasonably common in these areas.

The bushbuck, the smallest of this group, is found over much of Africa where there is enough cover for it to hide in. The sitatunga is a marsh dweller. It has a shaggy oily coat, and long broad hoofs which spread its weight as it walks on mud. When alarmed, it may submerge with only its nose above water.

CHOOSY FEEDERS

The various species of spiral-horned antelope have adapted to most of the main kinds of habitat in Africa. One skill they all have is the ability to pick out the small amounts of high-quality food from the surrounding poorer-quality plants. They are browsers rather than grazers. They take fruits, seed pods, flowers, leaves, bark and tubers. Grass is a minor part of the diet of all species.

SHY ANTELOPE

Even the big herds of open-country eland are shy. The smaller antelopes of this group are mainly active during the night or at dawn and dusk. Because of this, the thick vegetation in which many live, and their shyness, they are rarely seen even where they are common.

▼ The eland and its relatives have body markings, manes, beards and dewlaps that play a part in signalling.

DUIKERS

A tiny antelope creeps quietly through the bushes at the edge of a forest. It is sniffing the air, listening, and peering in front of it. A little way ahead, hidden in some undergrowth, is a bird's nest with three chicks in it. The antelope, a duiker, can sense the nest and moves closer. It finds the nest, little more than a scrape in the ground. With a sudden grab, it lifts one of the chicks and begins to eat it.

Duikers are small or medium-sized antelopes. All live in Africa south of the Sahara.

The name duiker is an Afrikaans word meaning diver. It describes the way the animals plunge into the undergrowth when disturbed. Duikers have short front legs, longer hind legs, and arched bodies. This is a good shape for slipping through dense thickets.

The sexes are usually similar, although females may be slightly the larger. Both have short conical horns, although these are sometimes missing in females, and are often hidden in a tuft of hair on top of the head. Most species are reddish or grayish brown, but some have patterned coats. The Yellow-backed duiker has a yellowish crest along its rump. This hair becomes erect when the animal is excited or alarmed. In some species, such as the Bay duiker or Jentink's duiker, the calf has a very different color and pattern from its parents.

HIDING IN THICKETS

Little is known about the life of many duiker species. Most live in and around dense forest thickets where they are difficult to watch, and they are shy and stay out of sight. The Common duiker is an exception. It is found in bush and savannah country, but even here it keeps within range of cover. Some species, such as the Blue duiker, are active during the day, but others are usually active at night.

Duikers are browsers, and pick high-quality food. They are unusual among antelopes in that, as well as eating the most nourishing pieces of plants, they also eat some meat. They have been seen to kill small animals, and often feed on carrion.

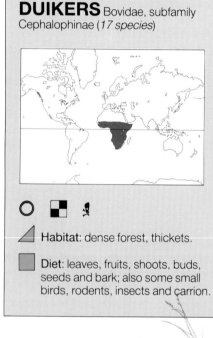

DUIKERS Bovidae, subfamily Cephalophinae (*17 species*)

○ ◼ ☠

◢ Habitat: dense forest, thickets.

◼ Diet: leaves, fruits, shoots, buds, seeds and bark; also some small birds, rodents, insects and carrion.

○ Breeding: 1 calf after pregnancy of 7½-8 months.

Size: smallest (Blue duiker): head-body 2ft, weight 11lb; largest (Yellow-backed duiker): head-body 5ft, weight 176lb.

Color: brown, gray, blackish, yellow, either plain or patterned, according to species.

Lifespan: 10 years.

Species mentioned in text:
Bay duiker (*Cephalophus dorsalis*)
Blue duiker (*C. monticola*)
Common duiker (*Sylvicapra grimmia*)
Jentink's duiker (*Cephalophus jentinki*)
Yellow-backed duiker (*C. sylvicultor*)

1

2

▼Some species of duiker A male Common duiker (1) dives for cover. The Yellow-backed duiker (2) with rump patch erect. The Zebra duiker (*Cephalophus zebra*) (3) of West Africa marks a bush with the gland in front of its eye. Jentink's duiker (4) has an unusual color pattern. Maxwell's duiker (*C. maxwelli*) (5) and the Red-flanked duiker (*C. rufilatus*) (6), here seen eating a bird, are two of the smallest.

▶Duikers often live in pairs. They are aggressive to outsiders. In some species the male and female mark one another gently with secretions from the glands just in front of the eyes.

DEVOTED DUIKERS?

Duikers do not live in herds. They are usually seen alone or in pairs. In the Blue duiker, males and females are known to form pairs that last for life, and this may be true of other species too. A pair of Blue duikers will live within its own small area, which is sometimes no more than the size of two football fields. Within this area the two animals may, for some of the time, travel, feed and sleep independently. There is no evidence that the male helps care for the young.

Even the mother duiker spends little time with the calf when it is young. The calf generally stays hidden in the undergrowth. Some of the smaller duikers are adult by 1 year old, and leave their parents behind to set up their own territory.

Duikers possess large scent glands beneath each eye. They put secretions from these glands on to bushes and trees, so marking their territory.

Duikers are hunted by many species, including humans, who find their meat good. The animals are easily dazzled at night with lights, making them quite easy to shoot or capture. Some species are also caught by driving them into nets. The rarest is Jentink's duiker. Discovered by science in 1892, probably only a few hundred remain in the wild, and even today few specimens exist in zoos.

WATERBUCK

Across the flooded plain comes a herd of 50 lechwe. They are running, spread out in a long line. They move in a series of bounds, jumping clear of the water, then plunging with a splash so that the water comes up to their bodies.

The waterbuck is well named for, along with its close relations, it lives mostly near water. This group of antelopes is found from the Nile basin to the south of Africa in habitats near water, and also in mountain grasslands. Apart from the waterbuck itself, the group includes four species of kob and lechwe, and three species of reedbuck. The little rhebok has sometimes been included in the group, but is probably only distantly related.

DAILY DRINKERS

The waterbuck lives from South Africa north to the Sudan and to Senegal in West Africa. It inhabits savannah country and also woodland, but always near water. Many individuals never enter the water, but they do need to drink regularly. This is because their food is rich and full of proteins, which must be balanced with a high intake of water. The waterbuck has a very greasy coat, which gives it a distinctive smell.

The waterbuck feeds on short and medium-length grasses, reeds and rushes. It also browses from bushes or trees. In the dry season it will wade in water and eat water plants. Because it eats a variety of plants, and lives in a rich habitat, a waterbuck may be able to find all it needs in a small area. In some parts of Africa a female may have a home range scarcely 600 sq yd. In other places, with poorer grazing, she might need up to 2½ sq miles. The home range is shared with five to eight

▼A small group of female waterbuck crop reeds at the water's edge. Although they keep within range of water, waterbuck are also found in dry savannah.

BREEDING GROUNDS

The kob antelope is found from West Africa to Uganda. Females live in groups of up to 50. Sometimes herds of over 1,000 animals gather to crop good areas of growing grass.

The adult males pack into areas that may have been used as breeding grounds for generations. These may measure only a few hundred yards long and wide. Within them the males each have a tiny circular area – perhaps less than 100ft in diameter – which they defend. Throughout the year the females visit the males at these sites to mate.

▲The Bohor reedbuck lives on the northern savannahs of Senegal, east to Sudan and south across Tanzania. The horns, with their strong forward hook, show this to be a male.

▶Gestures of grazing antelope The Southern reedbuck (1) has a white tail that shows as it flees. A waterbuck (2) stands tall to intimidate a rival. A kob (3) approaches a female with head held high. A rhebok (4) stands alert.

other females, which form a very loose herd. Often the animals move around separately.

Male waterbuck hold territories. Often these are alongside lakes or rivers where the grass is green. The holders allow females and young in, but usually push out strange males. Only a small proportion of adult male waterbuck are strong enough to hold territories, and thus are able to mate with females that come into them. Other males form bachelor herds.

CORRUGATED HORNS

The waterbuck and its relations have strong horns that curve forwards at the tips. In this group of antelopes, only males have horns. The horns have corrugations that run across their width, giving a good grip for the wrestling males. In the waterbuck the horns can be up to 3ft long.

The kob lives on gently rolling hills and low-lying flats close to permanent water, and grazes on shorter savannah grass than the waterbuck.

FEEDING IN FLOODS

The most aquatic of the waterbuck group are the lechwe. There are two species. The lechwe lives in Botswana, Zambia and Zaire. The Nile lechwe lives in the Sudan and western Ethiopia in the swamps by the Nile. Lechwe live on floodplains and move into grasslands when they are flooded. The puku is a related species that also likes wet conditions.

▼ Few animals like living in floodwater, but lechwe are adapted to living and feeding in these conditions.

The lechwe feeds on the leaves of grasses, and is often in the water. Usually it wades in depths of up to 8in covering its large hoofs, but may go so deep in search of food that the water covers its back.

Lechwe often come together in herds of hundreds, although the animals may take little notice of most of their companions. Large numbers of lechwe move together in response to rains and floods. In some parts of Africa dams have been built on rivers, spoiling the floodplains on which these animals rely. This has led to a drop in their numbers.

WHISTLERS

The Mountain reedbuck lives in mountain grasslands up to 16,500ft above sea level. It is found in three separate areas – the Cameroons, East Africa, and South Africa. It has a rather soft woolly coat. It feeds on poor-quality food. The females, each with their young, live in groups of two to six. They stay in the territory of a single male, or of several neighboring males. The animals give a whistling call which informs others of their presence and thereby spreads them out within an area.

The Southern reedbuck and Bohor reedbuck are lowland species which occur in southern and northern savannahs respectively. They seldom live far from water. They inhabit floodplains and the edges of flooded grasslands. They are particularly active at night, coming out from cover to

▼When fighting for territories and females, male reedbuck lock horns and try to twist each other's heads down to the ground.

feed on open lawns. They make whistling calls, and bounce as they move. Although they live near water they do not like wading. They eat grasses and the tender parts of reeds. In farming areas, they feed on young cereal crops. They are easy to approach, and rarely run far when they are alarmed.

HIGH-LIVING ANTELOPE

The rhebok lives in South Africa. It inhabits high mountain plateaux, or among the rocks and tangled plants of mountainsides. Where it feels safe, it may come down to grassy valleys. It is a small animal, weighing only about 50lb. Its coat is soft, gray and woolly. Only males have horns and these are short and grow straight up.

territories. They try to round up a harem of females for mating. But each male may hold a territory for only a few hours before the herd is on the move again, and his band of females merges into the mass.

MEAT-EATERS' BONUS
The old, the sick and the crippled bring up the rear of the Brindled gnu herds. Hyenas, lions and other meat-eaters follow the herds and make use of the easy opportunities to catch a substantial meal.

Migrating gnu have for generations used the same trails across the central Serengeti. They carry on through many obstacles, and swim rivers in their path. In the western Serengeti they spread out again into the good pasture. Just before the rains are due they begin to move back to the open grassland of the east. They reach it about the time the rains start, and are able to feed on the flush of new grass. At about the same time, and within a week or so, the calves are born.

DANGEROUS BABYHOOD
In some years the calves are born when the herds are still on the move. The herds are still attended by hunters and scavengers, so there can be no rest for the newborn. Very quickly they have to be up on their feet and able to move with the herd. Any that are too weak or too slow will perish. A mother gnu may bravely defend her baby, but she is unlikely to be a match for a group of hyenas or African Wild dogs. Many baby gnu do not survive the first few hours. In years when the rains are late and there is little grass on which the mothers can feed, three-quarters of the calves may die.

▶Even a flooded river does not stop the Brindled gnu on its annual migrations, although many animals may drown or die from injuries trying to cross.

ORYXES AND ADDAX

Vast expanses of sand dunes stretch in all directions. As if from nowhere come 15 white antelopes running in a group. They are addax. Their large hoofs stop them sinking and give a grip on the shifting sand. They are not running fast, but their long stride seems tireless. On they run, disappearing into the distance in search of a patch of grass.

Of all the large antelopes of Africa, the oryxes and their relations live in the driest regions. There are six species in this group.

DOOMED DESERT-DWELLERS?
The Arabian oryx used to live in the Arabian peninsula. It was saved from extinction only by being bred in captivity. The Scimitar-horned oryx lived over much of North Africa, but is now confined to a few areas and is much reduced in numbers. The addax, too,

is present in only a small part of its former range, which once included all the Sahara. These desert antelopes were once protected by living in such a harsh habitat, with few human inhabitants. When it became possible to drive vehicles into the desert, the animals became easy targets for people with guns, and they are now in great need of conservation.

It is too late to save one antelope of this group, the bluebuck. It lived in the south-west of Cape Colony in South Africa. It was an early casualty of the spread of Europeans with their farming and firearms. It was once abundant. The last one was shot in 1800. All that is left is a few stuffed specimens in museums.

POINTED HORNS
The long straight horns of the Arabian oryx grow to 27in long. The Scimitar-horned oryx has horns nearly twice as long, curving in an arc over its back.

ORYXES AND ADDAX Bovidae, tribe Hippotragini (6 species)

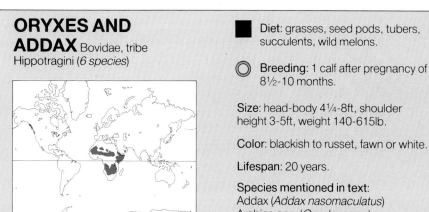

○ ■ ☠

● Habitat: dry grassland to desert, woodland edges.

■ Diet: grasses, seed pods, tubers, succulents, wild melons.

○ Breeding: 1 calf after pregnancy of 8½-10 months.

Size: head-body 4¼-8ft, shoulder height 3-5ft, weight 140-615lb.

Color: blackish to russet, fawn or white.

Lifespan: 20 years.

Species mentioned in text:
Addax (*Addax nasomaculatus*)
Arabian oryx (*Oryx leucoryx*)
Bluebuck (*Hippotragus leucophaeus*)
Gemsbok (*Oryx gazella*)
Roan antelope (*Hippotragus equinus*)
Sable antelope (*H. niger*)
Scimitar-horned oryx (*Oryx dammah*)

▼Arabian oryxes move to a new grazing area in the late afternoon. To survive the harsh environment, the herd needs to be highly organized. The dominant male is at the back, rounding up stragglers. A lower-ranking male leads.

◄Grazing antelope of dry regions The Roan antelope (1) and the Sable antelope (2) have long backward-curving horns. The Roan antelope is shown with its head dipping, the attitude used when giving way to superiors. The Sable antelope male pushes its horns forward to threaten another. The addax (3) has horns with a spiral twist. This male is sniffing to detect whether a female is ready to mate. The gemsbok (4) has long straight horns. The forward kick with the front leg is part of courtship.

WEARING WHITE

The desert oryxes have light-colored or even white coats. These help to reflect the heat of the Sun. Some oryxes live where daytime temperatures may reach 122°F, so they need to keep out as much heat as possible.

Oryxes and the addax have to be able to make the best of very sparse food resources. They eat coarse grasses, and may have to obtain all the water they need from this food. They produce little urine, and if shade is available, they use it when the Sun is strongest so that little water is wasted in sweating.

The addax seems to be able to sense the presence of a patch of food from far off, perhaps by smell. The gemsbok, in addition to grasses, feeds on seed pods, wild melons and cucumbers. It also eats various tubers and bulbs that it scratches from the soil. Some of these have a high water content.

MANED ANTELOPES

The Sable and Roan antelopes have long upright manes. They are found over much of the wooded savannah country in Africa south of the Sahara, but are not common animals. They usually stay within reach of water, and cannot stand drought in the same way as oryxes. They live in herds of 4 to 20, each herd including only a single adult male. Each herd may wander over an area of up to 120 sq miles during a

The horns end in sharp tips. Oryxes can use their horns effectively in defence. Some African tribes have used the horn tips as spearheads. The horns of all this group of antelopes have a strongly ringed appearance except at the sharp tip. Both males and females have horns.

The Beisa oryx of East Africa, which is known as the gemsbok in southern Africa, also has long, straight, pointed horns. This is the one species in the group which is still plentiful.

The biggest antelopes in the oryx group are the Sable and the Roan antelopes. These can stand 4¼ ft or more at the shoulders. The males are slightly larger than the females.

165

year, but in the breeding season usually stays within a circular area ¼ mile across. When the dry season comes, several herds often gather where grazing remains good, so up to 150 animals may be found together.

KEEPING APART

To a desert antelope such as oryx or addax there is little advantage in living as a big herd. Sometimes a few hundred may gather where rainstorms have started the desert vegetation growing strongly. More often, food is in short supply, and has to be searched for over a great distance. The desert antelopes live in tight little

▼Shade is precious in the desert. The Arabian oryx spends the hottest hours under trees if it can.

groups of less than 20 animals. These include females and young, and may have more than one adult male, although the "boss" male is likely to be the one to mate with all the females. The boss male threatens the lesser males quite often, but rarely comes to blows. The sharp horns of an oryx are almost too dangerous a weapon to use against companions.

Fighting has turned into a ritual. In an oryx "tournament" the animals in the herd run around in circles, sometimes breaking into a gallop. They pace stiffly as part of the ritual, and sometimes clash horns briefly.

A VALUABLE RESOURCE

The Arabian oryx was exterminated in the wild in 1972. But a few had been gathered together in captivity in

Phoenix, Arizona, and they have been bred to continue the species. This has been successful enough for animals to be returned to Arabia. In 1982 some were released in Oman, where their progress is being watched carefully. It seems a pity that the Arabian oryx should disappear, both for its own sake, and also because it could be a useful resource to people. The Arabian oryx is one of the few large animals that can make a living in these difficult desert conditions. They could be useful to people for milk, meat and hides. The ancient Egyptians tamed both the addax and oryxes.

In Arabia the midday temperature can vary by as much as 113°F from summer to winter. In a single summer's day, the temperature may vary by 68°F from noon to night. Rain may

▲ A male Sable antelope courts a female using a characteristic foreleg kick. In this species the mature males are black, the females brown.

▼ Gemsbok, with their long pointed horns, mingle with springbok to enjoy a drink at a waterhole in the semi-desert.

not fall for years. Plants remain dormant or in seed. Sand storms can blow for days on end. The Arabian oryx can cope with all these difficulties.

CLOSE COMPANIONS

The Arabian oryx is the smallest of its group of antelopes, and needs rela-

tively little food. Its hoofs are splayed and shovel-like, with a large surface to walk on sand. It is not a great runner, but it can walk for hours on end. It is quite usual for an oryx to travel 18 miles in a night to find food. Because it is small, the Arabian oryx can creep into the shade under the stunted thorn trees found in parts of the desert. Unlike some antelope, it feels comfortable close to others of its own herd. The animals share tiny patches of shade.

When they feed, the herd members spread out, sometimes 100yd apart. They keep looking up to see where the others are, and do not let themselves get left behind when the herd moves. An animal that gets separated from the herd can recognize and follow fresh tracks in the sand.

GAZELLES AND ANTELOPES

A herd of Thomson's gazelle are spread out on the plain. From the corner of its eye, each animal can see the black flank and twitching tail of the next. Suddenly a cheetah rushes at a gazelle at the herd's edge. The gazelle sees it and dodges away on a fast twisting run. The cheetah gives up. The rest of the herd have escaped in a series of bounds, their white rump hairs raised in alarm.

The gazelles and dwarf antelopes include 30 species. There are 18 species of gazelle which are medium-sized antelopes. Many live in dry, open country. Both sexes of most species have horns. Gazelles tend to be mainly sandy brown, fitting in well with their habitat.

The 12 species of dwarf antelope include some of the smallest hoofed mammals. Usually just the males are horned. They live in forest or in areas where there are dense thickets for cover. Gazelles are often found in herds, but dwarf antelopes live in pairs or as single animals.

TERRITORIAL MARKINGS

A dwarf antelope lives its life within a small area. It gets to know this area well and knows where to find food, shelter and escape routes from enemies. It protects this territory from other members of the same species, although a pair of antelopes may share the same territory.

Instead of fighting, dwarf antelopes keep away strangers by using scent signals. Scent glands on the hoofs leave a trail along pathways that a dwarf antelope regularly uses. Large glands in front of the eyes produce a

1

2

►Species of dwarf antelope and gazelle Klipspringer using dung to mark its territory (1). The rare beira (*Dorcatragus megalotis*) (2). The dibatag runs away with tail raised (3). Desert dwellers: Slender-horned gazelle (*Gazella leptoceros*) (4); Tibetan gazelle (5); Goitered gazelle (*G. subgutturosa*) (6); Dama gazelle, the largest gazelle (7). The oribi (8), the steenbuck (9) and Kirk's dikdik (*Madoqua kirkii*) (10), marking their territories. Royal antelope, the smallest (11). Blackbuck advancing in threat (12). A springbuck "pronks" in alarm (13).

GAZELLES AND ANTELOPES Bovidae, subfamily Antilopinae (*30 species*)

○ ■ ☠

◐ Habitat: from dense forest to desert and rocky outcrops.

■ Diet: young green leaves of bushes and grass, buds, fruit.

◎ Breeding: 1 young after pregnancy of 6-8 months.

Size: smallest (Royal antelope): head-body 1½ft, weight 4½lb; largest

(Dama gazelle): head-body 6ft, weight 185lb.

Color: light brown, golden, grayish or black, usually with lighter colored undersides.

Lifespan: 10-15 years.

Species mentioned in text:
Blackbuck (*Antilope cervicapra*)
Dama gazelle (*Gazella dama*)
Dibatag (*Ammodorcas clarkei*)
Dorcas gazelle (*Gazella dorcas*)
Gerenuk (*Litocranius walleri*)
Günther's dikdik (*Madoqua guentheri*)
Klipspringer (*Oreotragus oreotragus*)
Mongolian gazelle (*Procapra gutturosa*)
Oribi (*Ourebia ourebi*)
Pygmy antelope (*Neotragus batesi*)
Royal antelope (*N. pygmaeus*)
Springbuck (*Antidorcas marsupialis*)
Steenbuck (*Raphicerus campestris*)
Suni (*Neotragus moschatus*)
Thomson's gazelle (*Gazella thomsoni*)
Tibetan gazelle (*Procapra picticaudata*)

10

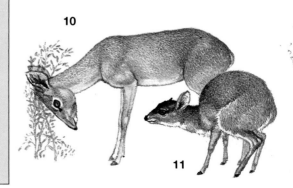

11

3

6

13

4

5

7

8

9

12

69

◄A male Kirk's dikdik has straight spiked horns. In front of the eye is the large gland used in scent marking. The snout is long, overhanging the bottom lip.

▼The male gerenuk, but not the female, has horns. Here a pair are courting. The male displays by turning the head sideways (1), then places scent from the gland in front of his eye on the female's rump (2). He taps the female with his foreleg (3) and sniffs her (4) before deciding whether she is ready to mate.

◀Springbuck flee from a predator. As they run they make sudden vertical "pronks" or leaps. This may confuse a predator and allows the gazelle to get a good view. A fold of skin that opens in alarm runs from mid-back to tail.

NOURISHING DIET

Most dwarf antelopes feed on the best parts of the vegetation around. They do not graze on old tough grass, but browse on bushes, taking the young green leaves. They eat buds, fruit and some fallen leaves and may pick tender new shoots of grass. Many of them thrive in places where the original vegetation has been disturbed by people and new vegetation is sprouting. This kind of food is packed full of nourishment.

SMALL AND CUNNING

The Royal antelope of West Africa competes with the chevrotains for title of smallest hoofed mammal. It is certainly the smallest horned mammal. It is shy and secretive, living in pairs in the thickest of forest. The Pygmy antelope of Central Africa and the suni of East Africa are almost as small and also live in forest with thick undergrowth.

All these have an arched back and short neck, which are useful for getting through the tangled growth. Female dwarf antelopes are usually a little larger than the males. In the folk-stories of several African countries dwarf antelopes have a reputation for cunning which makes up for their lack of size.

NOSE COOLERS

The three species of dikdik live in Africa in dry country with scrub and thorn bushes. They pair for life and are usually seen in pairs. Dikdiks, especially Günther's dikdik, have a long, rather fleshy nose. Water evaporates from the inside lining of this to help cool the animals down. Sometimes they pant through their noses. The name "dikdik" imitates the sound of the alarm cry they give as they dash away from danger.

The steenbuck and two species of grysbuck live in African woodland in the east and south. They are mostly solitary and have large home ranges. The oribi is the tallest of the dwarf antelopes, up to 27in high at the shoulders. It lives on grassy plains near water and, unusually for this group, eats grass.

ROCK JUMPER

The oddest dwarf antelope is the klipspringer. It lives in Africa in areas where there are dry rocky outcrops. It moves with a bouncing gait, and can arch its back to stand with all four feet together to balance on a tiny patch of level rock. It stands on the tips of its small round hoofs. Each hoof has a rubbery center and a hard outer ring, making a good pad to grip the rocks.

sticky secretion which the dwarf antelope wipes on to twigs and stems to help mark its territory. It also deposits dung and urine at particular sites, producing another scent marker. Both sexes add to the pile, a male adding his dung to a female's.

These markers usually repel possible intruders without a fight. One type of dwarf antelope, the dikdik, also threatens other males by "horning" (scratching at vegetation with its horns) and raising a crest of longer hairs on its head.

The hair is thick, bristly, light and hollow-shafted. It keeps the klipspringer warm and also makes a buffer to cushion the body from knocks on the rock.

CAMOUFLAGE IN THE OPEN

The gazelles live in a more open habitat than the dwarf antelopes. Most species have coats that are fawn above and pale beneath. This coloring helps their camouflage by disguising the shadows of an animal standing in the open. Some gazelles, such as the springbuck, have a black band along the side. This may also camouflage the animal by helping to break up the outline.

At close quarters such markings, and colored tails and rump patches, can work as signals among herd members. Gazelles have good hearing and eyesight. Smell seems less important to these animals.

The different species of gazelle live from southern Africa north to the Mediterranean, and across Arabia to India and China. In North Africa and Asia they live in dry or desert regions. Here their numbers are low and their populations are rather scattered.

MARCHING MILLIONS

Some gazelles live singly, but most form herds – although these are often little more than family groups. Large numbers of Thomson's gazelle live in East Africa, sometimes forming herds of up to 200. Thousands may come together when migrating to areas of fresher grass.

In the past the biggest herds were those of the springbuck of southern Africa on migration. Herds of millions formed, taking several days to pass a given point. Most were killed as a nuisance to farming. Now few springbuck are left, and their migrations are a thing of the past.

The blackbuck lives across the Indian subcontinent, in areas ranging from semi-desert to light woodland. It

▶ The typical stance of fighting gazelles is shown by these Thomson's gazelles. Their heads are as close to the ground as possible, with the strongly ringed horns interlocked.

▼ The gerenuk lives singly or in small groups. It is unique among antelopes in its ability to stand on its hind legs at full stretch to feed and to walk round a tree in this position.

is unique among gazelles in that the male is a different color from the female. A breeding male has a black back while the female has the usual gazelle brown. The male's color develops as it matures. It also has long spiral horns – the female has none.

A strong male blackbuck gathers a herd of up to 50 females and young in the breeding season. He threatens other males by lifting his head and tail and dropping his ears. Only rarely do threats lead to fights. The blackbuck is a strong runner, faster than almost any animal except the cheetah.

DESERT AND GRASSLAND

The Dorcas gazelle is a small species about 2ft tall, but able to run at almost 50mph. Youngsters are able to keep up with their mother within a week or two of birth. The Dorcas gazelle is found in scattered pockets from

Morocco to India in flat stony desert. It varies in size, coat color and horn shape through this range.

On the steppes and high plateaux of Central Asia live the Tibetan and Mongolian gazelles. Only the males of these species have horns. In Tibet, gazelles live as high as 18,000ft. Herds of Mongolian gazelles migrate across the grasslands to the summer pastures. Migrating herds may contain 8,000 animals. In the rutting season the males have swollen throats.

GIRAFFE GAZELLES

Two kinds of gazelle in Africa have very long necks and in their shape and habits are like giraffes. The dibatag is the smaller of the two, about 32in tall at the shoulders, and lives in Somalia. It lives in scrub desert with scattered thorn bushes and grass. The gerenuk is bigger, up to 3ft tall, with an even

longer neck than the dibatag. It lives from Ethiopia south to Tanzania in desert and dry bush savannah.

The feeding habits of these gazelles are like the giraffe's. They pick tender leaves and shoots from thorn bushes and other bushes and trees, also eating some flowers and fruits. They do not eat grass. Their narrow muzzles and mobile lips help them pick the tasty morsels from among the thorns.

To reach enough food, the gerenuk often stands on its hind legs and stretches upwards. Its diet varies between the rainy and dry seasons as the availability of different plant species changes too. In Tsavo National Park, Kenya, more than 80 plant species are eaten during the year. Some plants are evergreens, with thick hard leaves coated with a waxy layer that prevents evaporation of water. This allows the gerenuk to inhabit very dry areas.

▲A cheetah has pulled down and killed a Thomson's gazelle. Few other animals are fast enough to catch an adult gazelle unless it is taken completely unaware, but young babies are vulnerable.

GOATS AND SHEEP

High in the Rocky Mountains two horned male sheep face one another. These rams stand a little apart, lower their heads slowly, then rush at each other. They crash into one another head-on, with an enormous bang that echoes round the mountain. They back away, then launch themselves again, banging their enormous horns together with a jarring crash. Again and again they do this, until one accepts defeat and retreats. Neither animal has been injured by this head-banging.

Although it is easy to tell a domestic sheep from a domestic goat, it is not always so easy with the wild species. Females may be very similar. Males show more differences. In goats the males have chin beards and usually smell strongly. Goats have anal scent glands, and the males may also spray themselves with urine. Males have sharp scimitar-shaped horns with knobbed ridges. Goats have long flat tails with a bare underside.

Male sheep do not have chin beards but may have a throat mane. Sheep, unlike goats, have scent glands between the toes, in the groin and in front of the eye, but no anal gland. Male sheep (rams) do not have the offensive smell of goats. Rams have large, rather blunt horns which curl in a spiral. Sheep have short tails and often have a rump patch.

AGILE CLIMBERS

Goats and sheep tend to live in different types of country. Goats specialize in cliffs, while sheep live in the open, rolling dry lands close to cliffs. But both types of animal include agile climbers. This group of animals has made use of some of the most difficult, dangerous land, and eats

GOATS AND SHEEP
Bovidae; tribes Caprini, Rupicaprini, Saigini (*24 species*)

○ ■ ☠

◣ **Habitat**: often steep terrain, from hot desert and moist jungle to snowy wastes.

▢ **Diet**: grasses, leaves and bark of shrubs, other plants.

◎ **Breeding**: 1 or 2 lambs or kids after pregnancy of 150-180 days.

Size: head-tail 3½-6½ft, weight 55-300lb.

Color: usually shade of brown, some blackish, white or golden.

Lifespan: 10-15 years.

Species mentioned in text:
Argalis (*Ovis ammon*)
Barbary sheep (*Ammotragus lervia*)
Bighorn sheep (*Ovis canadensis*)
Chamois (*Rupicapra rupicapra*)
Chiru (*Pantholops hodgsoni*)
Goral (*Nemorhaedus goral*)
Himalayan tahr (*Hemitragus jemlahicus*)
Ibex (*Capra ibex*)
Japanese serow (*Capricornis crispus*)
Mainland serow (*C. sumatrensis*)
Markhor (*Capra falconeri*)
Mouflon (*Ovis musimon*)
Mountain goat (*Oreamnos americanus*)
Saiga (*Saiga tatarica*)
Snow sheep (*Ovis nivicola*)
Thinhorn, or Dall, sheep (*O. dalli*)
Wild goat (*Capra aegagrus*)

▼ ▶ **Goats or goat antelopes** Goral (1), Japanese serow (6), chamois (7) and Mountain goat (9) have both sexes similar in size and looks. They have short, sharp horns. In many other species males are bigger and have beards and manes. A male Himalayan tahr (2) may weigh 235lb. The male urial (*Ovis orientalis*) (3) and male Barbary sheep(4) have spiral horns and a throat mane. The Wild goat (5) and ibex (8) have long curved horns. The argalis (10) has huge horns.

▲The Mountain goat has a thick coat with a layer of fat beneath to keep it warm in bitter cold.

some of the toughest plants.

There are 24 species of goats and their relatives. Some are rather odd looking, such as the serows, chamois, goral and Mountain goat. More "typical" in appearance are the tahrs and the true sheep and goats. There are also two bigger species of "giant goat," the Musk ox and takin (pages 80-81).

SIMILAR TO GAZELLES

On the borderline between the goats and the gazelles are two species, the saiga and the chiru. The chiru lives high in the plateaux of Tibet in small groups. Only the male has horns. The same is true of the saiga. This animal once lived across a wide area, from Poland to Central Asia, in large numbers. People killed so many Saiga that the species nearly became extinct. But it is now protected and out of danger.

The saiga migrates long distances across the steppes, and huge herds form at this time. The saiga is a fast runner, reaching speeds of up to 35mph. It has a large bulbous nose, which helps to warm the air it breathes. It also has a good sense of smell.

The chiru and saiga look and behave like gazelles, and this is what the majority of scientists now believe them to be.

STURDY AND SURE-FOOTED

The serows are rather clumsy and slow compared to goats, but they are sure-footed as they go up and down the steep rocky slopes of their home areas. They live where there is shrub or tree cover. The Mainland serow is found from India south to Sumatra and east to southern China. The Japanese serow is found in much cooler conditions, including snow, in Japan and Taiwan. Both species of serow are browsers.

The goral is a smaller but similarly shaped animal, with a slightly shaggy coat, living over a wide area of Asia from India to Thailand and north to Siberia. It usually lives in dry climates on very steep cliffs. The goral climbs and jumps well. It feeds morning and evening and may rest on a sunny ledge for much of the day.

GOATS OF THE SNOWS

The Mountain goat lives in western North America. It is found in rocky areas high in the mountains, often above the tree line. It climbs steep cliffs and along the edges of large glaciers. The Mountain goat is sure-footed, but moves slowly most of the time. It eats grasses and lichens and

▼A domestic sheep cleans her newborn lamb. The domestic sheep's dense woolly coat is not found in wild sheep.

▶A group of Dall sheep stay alert as they rest and chew the cud on a pasture high in the mountains of Alaska.

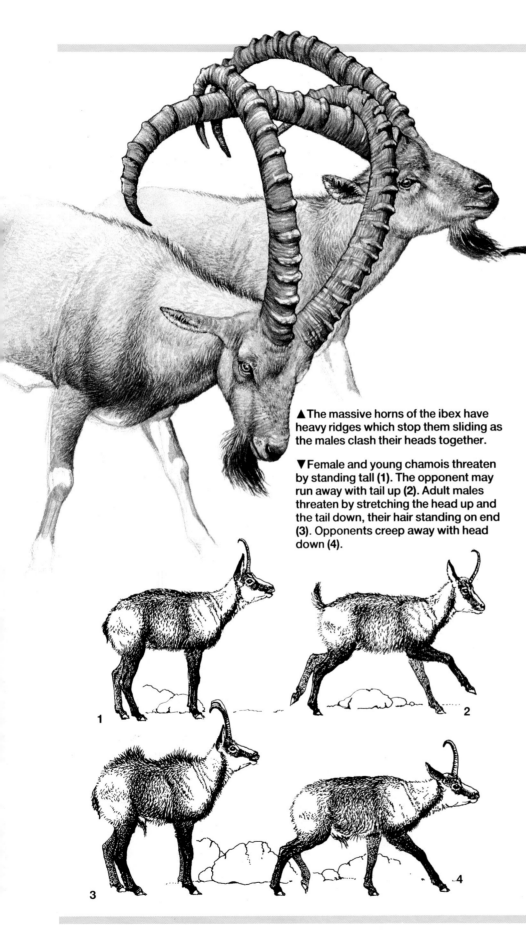

▲ The massive horns of the ibex have heavy ridges which stop them sliding as the males clash their heads together.

▼ Female and young chamois threaten by standing tall (1). The opponent may run away with tail up (2). Adult males threaten by stretching the head up and the tail down, their hair standing on end (3). Opponents creep away with head down (4).

may browse on bushes. Because it lives where the snowfall is heavy, the Mountain goat has developed a stocky build and a thick white woolly coat. The thick legs look as though they are in pantaloons. They have very large hoofs with hard rims and softer centers which give a good grip.

The most nimble of the "near-goats" is the chamois. This lives in the cold snowy mountains of Europe and Asia Minor. The chamois comes down the mountain to spend winter in woodland, where it feeds on buds, young tree shoots, lichens and small grass patches. Summer is spent high on the mountains, feeding on grasses and herbs.

The chamois can make use of tiny areas of good footing to leap up and down the most unlikely looking rock faces at high speed. A whole herd will throw itself down a cliff which, to a human, looks almost impossible to climb. Even the babies are agile. The kids are born in May and June on rocky, inaccessible areas. They can follow their mother almost immediately, and within a week are at home jumping about the crags.

Old male chamois are usually solitary, but females live in herds of up to 30. Males fight viciously during the breeding season, using their hooked horns. Unless a loser submits by flattening himself on the ground, he may be gored to death.

BEARDLESS GOATS

Tahrs live on tree-covered mountain slopes and cliffs. They have no beards and a naked muzzle. The horns are rather flattened. Males do not smell like true goats. The Himalayan tahr has a thick shaggy mane around its shoulders. Other species live in southern India and in Oman. Tahrs are wary animals. A few animals in the herd of up to 40 are always on watch.

The Barbary sheep is native to North Africa. It is found in mountainous areas which are barren and rocky.

It seems able to survive without drinking water. In some respects it is more like a goat than a sheep, and it has a long flat tail. But it fights like a sheep, by clashing heads together. The horns of big male Barbary sheep are large. Males also have a body weight twice that of females.

TRUE GOATS
Six species of wild goat are found from the Pyrenees to central China. They are mainly browsers, but also graze. The ibex exists in a number of slightly different forms in various mountain ranges. The horns of the Siberian ibex are the longest, growing to 4½ft.

The horns of the markhor grow equally long, but they are twisted. The markhor lives in the mountains of Central Asia and is the largest goat. Males can be 40in high at the shoulders and weigh 240lb.

The Wild goat from which our domestic animals were bred still exists in western Asia and the eastern Mediterranean. It was first tamed in western Asia approximately 8,500 years ago. Its eating habits prevent the regrowth of trees in some parts of the world.

GRAZING IN HERDS
Six species of sheep live from the Mediterranean across Asia to Siberia and in the west of North America. Most live by grazing. They are herd animals which rely on keeping together to escape enemies.

The mouflon is a small sheep found in Asia Minor and the Mediterranean. This is probably the species from which tame sheep were bred. Some of the old-fashioned breeds like the St Kilda sheep still have much the same size and coloring as the mouflon. It is chestnut brown with a light saddle. The mouflon is hardy and can live in cold or desert habitats. Domestic sheep have been bred for various characters, including a long woolly fleece, and there are now many breeds, mostly with white coats.

MOUNTAIN SHEEP
The largest wild sheep is the argalis, which lives in cold desert and mountain habitats in Central Asia. It has a light brown coat with a white rump patch. A male can be 4ft tall and weigh 400lb. The horns curve in a spiral and can be 6¼ft long and 20in round, weighing up to 50lb.

The Snow sheep lives in the mountains and Arctic wastes of Siberia. The Thinhorn sheep lives in similar conditions in Alaska and western Canada. The Bighorn lives in mountainous and dry desert areas from Canada to Mexico. In the Rocky Mountains it occupies the same areas as the Mountain goat, yet the two species have very different life-styles. The Bighorn is migratory, moving seasonally across wide areas of woodland, while the Mountain goat ruthlessly defends a territory when food becomes scarce on the cliffs in deep snow.

▼The chamois moves lower down the slopes in the winter, but is still likely to encounter snow in its mountain home. These animals are at 7,000ft in the Swiss Alps.

MUSK OX AND TAKIN

It is the middle of the Arctic winter. Although it is permanent night, light from the Moon comes out of a clear sky. The wind whips across the open landscape and the temperature is −70°F, colder than most food freezers. A small group of Musk ox are feeding, scraping with their hoofs at the thin snow cover. Under the snow are frozen lichens and other low plants. The animals eat them silently, taking no notice of the bitter cold.

The Musk ox is more closely related to goats than to cattle. Its closest relative of all is the takin. The takin lives in a temperate climate in China and Burma. The Musk ox, though, lives in the bitterest climate of any mammal, in the high Arctic of North America and on Greenland.

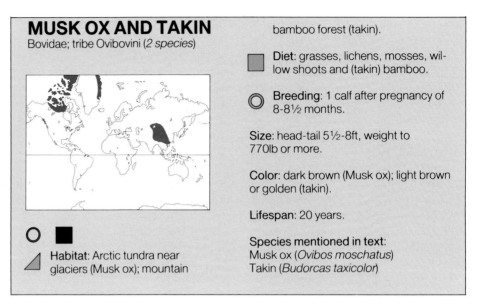

MUSK OX AND TAKIN
Bovidae; tribe Ovibovini (*2 species*)

bamboo forest (takin).

Diet: grasses, lichens, mosses, willow shoots and (takin) bamboo.

Breeding: 1 calf after pregnancy of 8-8½ months.

Size: head-tail 5½-8ft, weight to 770lb or more.

Color: dark brown (Musk ox); light brown or golden (takin).

Lifespan: 20 years.

Species mentioned in text:
Musk ox (*Ovibos moschatus*)
Takin (*Budorcas taxicolor*)

○ ■

△ **Habitat:** Arctic tundra near glaciers (Musk ox); mountain

GROUND-LENGTH COAT
The coat of the Musk ox has two parts. The inner coat is soft, very fine and light. It is so thick that cold does not penetrate. Growing through this there are long guard hairs, forming a long overcoat which reaches nearly to the ground. Individual hairs may be as

much as 2ft or more long.

This outer coat can shrug off snow and rain and keeps out the wind. It takes the wear and tear, so that the Musk ox's warmth-giving inner coat stays in good working order. In early summer the inner coat is shed and replaced. The compact shape of the

Musk ox, with its short legs, neck and tail, also helps it retain heat.

The Musk ox's coat makes it look bulky, but it runs well. It is able to turn at surprising speed.

SEX DIFFERENCES

During the rut the bulls produce a secretion from glands just in front of the eye. They fight each other by running together and banging their heads, making a noise that can be heard more than a mile away.

The horns form a helmet over the top of the skull, nearly joining in the midline. They curve down, then out. The part of the horn on top of the skull may be over 8in thick in a big male, making a cushion for crashing skulls.

Females have horns too, but they are smaller. A female Musk ox weighs little over half the male's weight.

PROTECTIVE BUNCH

The Musk ox is a social animal and lives in herds. Often a herd has only about 12 animals, but there can be as many as 100. They stay in a close bunch for protection against enemies and bad weather.

In summer the Musk ox feeds well on the tundra, on grasses and dwarf shrubs. It builds up a layer of fat which protects it from cold and starvation in the winter. The Musk ox can survive only in areas where the snow cover is not deep enough to bury the food out of reach.

The Musk ox calf is born in May. Even then the temperature can be well below freezing, and it has a thick woolly coat. It is about 18in tall and after an hour can follow its mother.

▲The takin lives in the highest bamboo forests and rhododendron thickets. It eats grasses and bamboo shoots.

▼The Musk ox huddles in steaming groups for warmth. They are the only large mammals to spend all year on the treeless Arctic tundra.

▼The takin needs much less fur than the Musk ox, but it has a heavy body and large hoofs. Some races are golden in color. Most are light colored, with dark faces in bulls. The takin's skin secretes a strong-smelling grease.

▼When threatened by enemies, a herd of Musk ox form a line facing them, or a circle with the calves in the middle. Big males dash out and jab with their horns. This is a good defence against wolves, but not against people with guns.

KANGAROOS

KANGAROOS Macropodidae and Potoroidae
(60 species)

weight up to 200lb. In larger species, males bigger than females.

Color: mainly shades of brown or gray; some have contrasting facial markings or stripes on body or tail.

Lifespan: up to 20 years in the wild, 28 years in captivity.

Species mentioned in text:
Eastern gray kangaroo (*Macropus giganteus*)
Lumholtz's tree kangaroo (*Dendrolagus lumholtzi*)
Musky rat kangaroo (*Hypsiprymnodon moschatus*)
Red kangaroo (*Macropus rufus*)
Wallaroo or euro or Hili kangaroo (*M. robustus*)
Western gray kangaroo (*M. fuliginosus*)

○ ■ ☹

◁ **Habitat:** inland plain and semi-desert to tropical rain forest and hills.

■ **Diet:** grasses, other low-growing plants, shoots of bushes.

◎ **Breeding:** 1 joey after pregnancy of 27-36 days, plus a period in the pouch of 5-11 months.

Size: smallest (Musky rat kangaroo): head-body 11in, plus 5½in tail, weight 1lb; largest (Red kangaroo): head-body 5½ft, plus 3½ft tail,

Two gray kangaroos sip water from the edge of a small river. They turn and move slowly up the bank. Suddenly alarmed by a flock of birds landing to drink, they begin to move fast. They thump their back legs down hard and spring into the air. Each bound sends them higher and faster, until they are in full flight, travelling at 30mph in jumps 13ft long.

The kangaroo family is large. There are about 50 species. The larger species are called kangaroos, the smaller ones wallabies. There are also 10 small species known as rat kangaroos and bettongs. Kangaroos are marsupials, animals of which the female has a pouch where she keeps the young.

HOW THEY HOP
The whole group have long back legs and travel fast by hopping. The long tail works as a counterbalance to the weight of the body as they jump. When moving slowly, kangaroos may use the front legs and tail as a tripod to

▼ ▶ **Larger kangaroos and wallabies**
The Bridled nailtail wallaby (*Onychogalea frenata*) **(1)** lives in open country. The wallaroo **(2)** is widely distributed in Australia. The rabbit-sized quokka (*Setonix brachyurus*) lives in West Australia **(3)**.The Red-legged pademelon (*Thylogale stigmatica*) **(4)** and the Yellow-footed rock wallaby (*Petrogale xanthopus*) **(5)** are mainly nocturnal. The Gray forest wallaby (*Dorcopsis veterum*) lives in New Guinea **(6)**

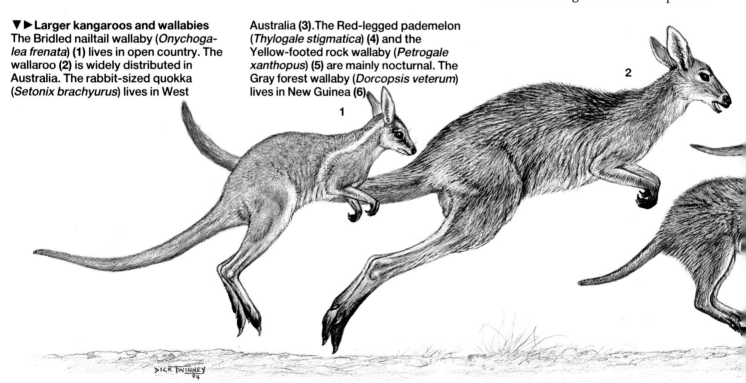

DICK TWINNEY 84

take the weight, then they swing the back legs forward. Next, the back legs support them while they move their front legs and tail.

The long and narrow back foot has four toes. Two of these are large. The other two are small and joined together. They make a kind of comb which a kangaroo uses to keep its fur in good condition.

Kangaroos are plant eaters. They have good chewing teeth in the back of the mouth. As with elephants, their teeth move forwards as they grow older and the front teeth wear down. Kangaroos have large stomachs where there are bacteria which help break down the tough plant food. Most kangaroo species feed at night.

WIDE VARIETY

The big Red kangaroo lives mainly on open plains. It has a thick woolly coat which helps keep out both heat and cold. During the heat of the day it rests in the shade of a bush, feeding when it is cooler. It needs little water, but even in a dry brown landscape it will find newly sprouting grasses and herbs to eat.

▲Male, female and joey (young in pouch) of the largest marsupial, the Red kangaroo, which lives in grassland.

▼Kangaroos and wallabies, as plant-eaters, are the Australasian equivalents of African hoofed mammals.

3 4 5 6

83

Wallaroos are also desert animals and can exist on even poorer food. The Eastern and Western gray kangaroos live in wooded country. They eat the grasses that grow between trees and they need to drink more often.

Many smaller wallabies like to live in scrub or dense thickets. In some parts of Australia much of this kind of vegetation has been cleared for farms or ranches, and this is one reason why some of the small kangaroos have become rarer. Another reason is that they have suffered from introduced rabbits competing for food and living

▲The back legs of tree kangaroos are fairly short, but the tail is long and provides a good balancing limb.

▼Small- and medium-sized kangaroos
The Lesser forest wallaby (*Dorcopsulus macleayi*) (1) lives in New Guinea. The Musky rat kangaroo (2). Lumholtz's tree kangaroo (3). The Desert rat kangaroo (*Caloprymnus campestris*) (4). The rare boodie (*Bettongia lesueur*) (5) burrows. The Rufous hare wallaby (*Lagorchestes hirsutus*) (6). The Long-nosed potoroo (*Potorous tridactylus*) (7) and the Rufous bettong (*Aepyprymnus rufescens*) (9) are now rare. The Banded hare wallaby (*Lagostrophus fasciatus*) (8) survives only on coastal islands.

DICK TWINNEY
84

places, and also from being hunted by introduced cats and foxes.

Some kangaroos live in hot wet forests, such as are found in northern Australia and New Guinea. Tree kangaroos (such as Lumholtz's tree kangaroo) climb, and have some of the brightest colored fur among marsupials. They are agile, making big leaps from one tree to another. But on the ground they are rather slow and clumsy.

SOCIAL AND FAMILY LIFE
Many of the smaller kangaroos live alone, but the Red and grey kangaroos live in groups (called "mobs"). From 2 to 10 move around together, but larger numbers may come together where food is good. In the largest species the males may be twice the size of the females. In Red kangaroos and wallaroos the males and females are different colors.

Kangaroos, like other marsupials, have a very short pregnancy, about a month long. Even in the biggest kangaroos, the baby weighs less than $1/30$ ounce at birth. The baby has big arms and small legs. It crawls by itself to the mother's pouch, where it attaches to one of the four teats. Here it suckles and grows. After several months it takes trips outside the pouch. Even when it leaves the pouch completely (up to a year after birth) it nurses for a few months more.

85

PANDAS

A Giant panda pushes its way through the bamboo forest with a rolling walk. It is seeking the spot it found yesterday where tender new bamboo shoots were sprouting. It comes to the clearing it remembers, but the place is already occupied. Another panda is sitting there, chewing on the bamboo. The new arrival mutters a few growls, then sits in the clearing far from the other panda. It turns its back and starts to feed.

► The Giant panda looks very human as it sits with its body upright and puts food to its mouth with its front paws.

▼ As well as five toes, the Giant and the Red panda have an enlarged wrist bone that can be folded over to work as a "thumb." In this way pandas can grasp the bamboo shoots that they love to eat.

▼ Giant pandas look like bears, which genetic evidence shows are their closest relatives.

PANDAS Procyonidae
(2 species)

● ◪ ☠

◤ Habitat: mountain forest.

◪ Diet: bamboo shoots, grasses, bulbs, fruits, some insects, rodents and carrion.

○ Breeding: 1 or 2 cubs after pregnancy of 125-150 days (Giant panda); 1-4 young after pregnancy of 90-145 days (Red panda).

Size: Red panda: head-body 1½-2ft plus 16in tail, weight 6½-11lb; Giant panda: head-body 4-5ft plus 5in tail, weight 220-230lb.

Color: black and white (Giant panda); reddish chestnut, black below, light markings on face (Red panda).

Lifespan: 14-20 years.

Species mentioned in text:
Giant panda (*Ailuropoda melanoleuca*)
Red panda (*Ailurus fulgens*)

There are two species of panda. The Giant panda is black and white and bear-like. It lives in central and western China. The Lesser or Red panda has a beautiful reddish-chestnut back, darker below and on the legs, and lighter face markings. It is shaped like a chubby cat and has a long tail. It is found from Nepal to western China in mountain forests.

The Red panda was known to Western scientists a long time before the Giant panda was discovered in 1869. The Red panda finds most of its food in trees. It eats bamboo, fruit, acorns, lichens, roots and some small animals. Like the Giant panda it has an extra wrist bone that can be used as a thumb, though not so well. It is nocturnal and lives a solitary life.

BAMBOO DIET
The Giant panda, like the Red panda, belongs to the group of mammals called the Carnivores. Most of its relations, such as lions and wolves, eat meat most of the time, but the Giant panda has adapted to a life of feeding mainly on bamboo. It feeds on both the shoots and the roots but, whenever available, it prefers the leaves and slender stems. The Giant panda's cheek teeth are specialized for slicing and crushing food, and it can cope with stems up to about 1½in in diameter. It also eats some bulbs and tubers of other plants, grasses and some small animals.

Although bamboo forms the main part of its diet, the Giant panda's gut is not especially efficient at digesting it, and much of the bamboo passes straight through the body. To get enough nourishment the Giant panda may spend as much as 12 hours a day feeding.

MATING AND RAISING YOUNG
The Giant panda is largely solitary in the wild. There are scent glands beneath its tail, and it rubs these against large objects in the surround-

▲ The Red panda has big whiskers. With its long tail, short legs and sharp claws it is well built for climbing.

ings. This probably marks the territory and keeps away other pandas of the same sex. Even males and females are little interested in one another except in the brief mating season in the spring. Male and female find one another by scent and sound. After a brief mating they separate again.

A pair of cubs, sometimes three, are born in a sheltered den. Normally only one survives. Cubs are small, blind and helpless, weighing only about 3½ ounces. Their eyes do not open for 6 or 7 weeks, and the cub cannot follow its mother until it is about 3 months old. It is weaned at 6 months. At a year old it may move off to live independently. Female Giant pandas are able to reproduce when about 4 or 5 years old, but the males are probably not fully mature until about 6 or 7 years old. Full-grown males are about 10 per cent larger and 20 per cent heavier than females.

PANDA PROBLEMS
The Giant panda is a rare animal. Fewer than 1,000 survive in the wild, and zoo breeding has not been successful. It has probably always been a rare animal, because its way of life confines it to a limited area – bamboo forests at heights of 8,600 to 11,550ft. Although it is protected and lives in remote areas, its populations are so small that it is likely to die out.

One of the panda's chief foods is a bamboo which flowers only about every 100 years and then dies back. This has happened recently in some parts of the panda's range. Yet as a species the Giant panda must have survived these flowerings many times before and could do so again.

SEA COWS

A pair of nostrils break the surface of the murky water, and then their owner dives again. It is over a minute before the nostrils appear once more. Below the water the dugong is busy, digging up the roots of sea-grass and chewing them. Just out from the shore other dugongs are feeding on the sea-grass beds, but they take little notice of one another. Yet when a small motor boat chugs past, several raise their heads to peer at the intruder.

There are four living species of sea cow. The dugong is found in coastal shallows from the South-west Pacific to the coast of East Africa. The West Indian manatee lives in coastal waters, estuaries and rivers from Florida, through the Caribbean, to Brazil. The West African manatee lives in similar habitats down the coast of tropical West Africa, while the Amazonian manatee lives entirely in river water.

DIET OF SEA PLANTS
Sea cows live in tropical waters where sea-grasses (similar to the grasses that grow on land) grow in the shallows.

They do not eat seaweed. Much of the food-value of the sea-grasses is in their underground stems, and sea cows dig these up and eat them as well as the leaves. To deal with their tough plant

▼Species of sea cow Steller's sea cow (*Hydrodamalis gigas*) is extinct (1). 26½ft long and weighing 6 tons, it lived in the cold Northern Pacific. It was discovered by Europeans in 1741 and all were killed by 1768. The Amazonian manatee (2) feeds on floating vegetation. The West African manatee (3) has strong bristles and mobile lips typical of sea cows. A West Indian manatee (4) carries food in its flippers. The dugong (5) tail is forked, not round as in manatees.

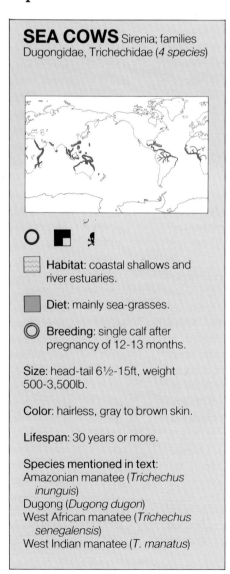

SEA COWS Sirenia; families Dugongidae, Trichechidae (*4 species*)

○ ■ ⚔

〰 Habitat: coastal shallows and river estuaries.

▪ Diet: mainly sea-grasses.

◎ Breeding: single calf after pregnancy of 12-13 months.

Size: head-tail 6½-15ft, weight 500-3,500lb.

Color: hairless, gray to brown skin.

Lifespan: 30 years or more.

Species mentioned in text:
Amazonian manatee (*Trichechus inunguis*)
Dugong (*Dugong dugon*)
West African manatee (*Trichechus senegalensis*)
West Indian manatee (*T. manatus*)

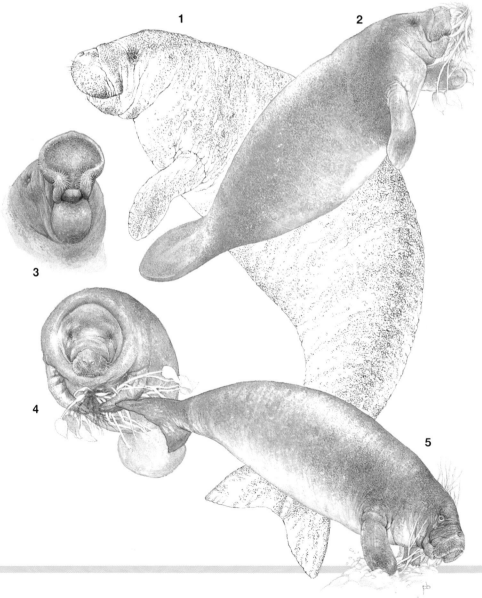

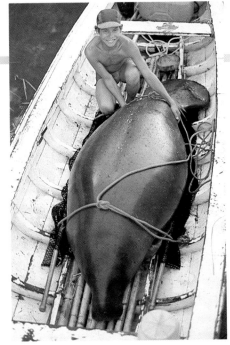

▲A manatee is captured for transport to another area where it will be used to control water-weed in a dammed lake.

▼Body contact such as "kissing" (1) takes place between manatees. The only constant social group seems to be mother and young (2). Manatees may rest on the bottom on their back (3). They use rubbing posts (4) for scent marking.

food, manatees have intestines more than 145ft long.

Sea cows are streamlined for swimming, although they rarely swim fast. They do not have back legs while their forelimbs are flippers which help in steering, but are flexible enough to help gather food. The tail is horizontally flattened for swimming. The fat that sea cows store under their skin helps keep them warm, but they still avoid temperatures below 68°F. Sea cows' bodies work at only a third of the rate of most mammals.

Manatees have teeth which grow forward, wear away and are replaced from the back. This makes up for the toughness of their food. Dugong teeth are even odder. The dugong has a pair of short tusks, then a few peg-like teeth at the back of the mouth. The work of chewing is done by horny plates covering the palate and the front of the lower jaw.

The dugong rakes in food using its mobile top lip and bristles. The muscles contract in waves to extract food items and pass them back to the grinding jaws. The dugong digs most of its food from the floor of the shallows with its snout, earning it the alternative name of sea-pig.

The West Indian and African manatees also forage on the bottom. The Amazonian manatee usually feeds on plants which grow on the surface.

GENTLE HELPERS

Dugongs sometimes occur in herds of several hundred, and manatees in herds of a dozen. Yet these animals seem to be basically solitary, having no herd organization. They breed slowly – a female dugong may live 50 years but will produce only six young. The baby is suckled for nearly 2 years. It is 8 years old before it breeds.

Sea cows are gentle creatures. For years people have killed them for their meat, oil and hide. Now they are in demand to help clear waterways and lakes of vegetation.

GLOSSARY

Adaptation Any feature of an animal that fits it to live in its surroundings. It can be something about the way the animal is built – such as the long neck of the giraffe – or it can be something it does.

Adult A fully developed animal which is mature and capable of breeding.

Alpine Belonging to the Alps or any other high mountains. Usually used for heights that are over 4,700ft.

Anal gland A gland opening into the anus or close beside it. Many mammals have scent glands close to the anus or the tail.

Aquatic Living in water.

Behavior The way that an animal acts.

Brindled Having a coat made up of a mixture of light and dark flecks, usually dark on a gray background.

Browser A plant-eater that feeds on shoots, leaves and bark of shrubs and trees as opposed to grass.

Carnivore Any animal that feeds on other animals.

Cellulose The chemical which makes up the cell walls of all green plants. It is very tough and fibrous. Mammals cannot digest it except with the help of bacteria that live in their gut.

Cheek-teeth Teeth that lie behind the canines (fang-teeth) of mammals. The cheek-teeth of plant-eaters are usually broad and ridged to give a grinding surface to break up the tough food.

Coniferous Describes trees such as fir and pine which bear cones. Their leaves are thin needles, and most coniferous trees bear leaves all year.

Cud Food which is brought back into the mouth, by plant-eaters such as cows, from the first part of the stomach to be given a thorough chewing. It is then sent to the second part of the stomach.

Deciduous Describes trees such as oak and beech, which have leaves which are shed all together at a particular season.

Den A shelter made or used by an animal for sleeping or for looking after young.

Desert An area with low or no rainfall. Deserts may have thin scrub or grassland vegetation, or none at all.

Digit A finger or toe.

Display An easily noticed pattern of behavior that gives information to other animals. It may be made up of behavior that can be seen or heard. Greeting, courtship or threatening may involve displays between animals.

Distribution The whole area in the world in which a species is found.

Dominant Describes an animal which is the "boss" of a group, or of a higher rank within a group than another animal. A dominant animal may get the pick of food, mates, shelters, and so on.

Dormant Inactive, resting, as a bear for example, when it sleeps for much of the winter.

Dorsal Towards or on the back of an animal rather than its belly.

Family In classification of animals, a group of species which share many features in common and are thought to be related, such as all the pigs in the Pig Family, Suidae.

Feces Waste matter passed from the gut through the anus. Also known as droppings.

Grazer A plant-eater that feeds on grasses and gathers its food from the ground, rather than from tall plants.

Guard hairs Long hairs which grow in an animal's coat and form a barrier on the outside to protect the shorter coat below. In many species the guard hairs are stiffer than the undercoat.

Habitat The surroundings in which an animal lives, including the plant life, other animals, physical surroundings and climate.

Harem A breeding group of two or more females that are attended by just one male.

Herbivore Any animal that feeds on plants.

Hibernation A winter period in which an animal is inactive. In true hibernation the body processes slow down and the body temperature drops.

Home range The area in which an individual animal normally lives, whether or not the area is defended against other animals of the same kind.

Introduced Describes an animal which has been brought by humans to a particular area. Introduction may be deliberate or accidental.

Juvenile A young animal which is no longer a baby, but is not yet fully adult. Sometimes referred to as an adolescent.

Mammal A backboned animal in which the babies feed on milk from their mother's milk (mammary) glands.

Mangrove forest A type of forest which develops on the muddy shores of the sea in deltas and estuaries in the tropics.

Marine Living in the sea.

Migration Movement, usually seasonal, from one region or climate to another for the purpose of feeding or breeding.

Niche The particular way of life of a species in a certain habitat.

Nocturnal Describes an animal that is active during the hours of darkness.

Omnivore Describes an animal which feeds on all types of food, including both plants and animals.

Opposable Can be put opposite one another. Our thumb is opposable to our fingers.

Palmate Like the palm of a hand. Describes the antler shape of deer such as fallow deer.

Pampas Grassland of a type found in South America, especially Argentina, where there are vast treeless plains.

Pouch A flap of skin on the belly of female marsupials that covers the teats. It may be a simple open structure, as in many marsupial carnivores, or a closed pocket, as in kangaroos.

Prairie A type of open grassland found in North America.

Predator Any animal that hunts, catches and kills another animal for food.

Prehensile Able to grasp, as the tail of some monkeys or the lips of some plant-eaters.

Proboscis A large nose such as that of an elephant.

Race Those animals of a species, living within a certain geographical area, that have characteristics in common that mark them off from the rest of the species.

Rain forest A type of forest found in the tropics and subtropics that has a heavy rainfall throughout the year. Such forests usually contain a large number of different species.

Range The area in which an animal lives; *see* home range.

Rodent A mammal of the group that includes rats, mice, squirrels and guinea pigs. They have large gnawing teeth at the front of the mouth, and chewing teeth in the cheeks.

Rut The mating season, particularly in discussing mammals such as deer.

Savannah A type of grassland found in the tropical parts of Africa, America and Australia. Savannahs have a seasonal rainfall, and are typically on plains or plateaus. They often have scattered trees.

Scent glands Special organs or areas of skin which produce chemicals which can be smelled, especially by animals of the same kind. The scent may not be detectable by humans, and may not be pleasant if it is, but such scents are an important means of communication in mammals.

Scrub A type of vegetation in which shrubs are the main part. It occurs naturally in some dry areas, or can be produced by human destruction of forest.

Sedentary Staying in the same area, not wandering or migrating.

Shrub A woody plant, low-growing or shorter than a tree, with several stems rather than a single trunk.

Solitary Living alone, not in a group.

Sounder The name for a group of pigs.

Species A group of animals of the same structure which are able to breed with one another. For example, the lion is one species, the tiger another.

Taiga The type of forest found in the cold lands of the north. It is made up of coniferous trees with open rocky and boggy patches in between.

Terrestrial Living on land.

Territory An area which is defended by an animal, or group of animals, against other animals of the same species.

Tundra Barren treeless lands found in the far north of Europe, Asia and North America. Tundra may also be found on mountain tops. The vegetation consists of low shrubs, low-growing perennials, mosses and lichens.

Underfur The thick soft fur lying beneath the guard hairs of some mammals.

Ungulate A mammal which has hoofs.

Velvet The skin covered with soft fur which covers a growing antler in deer.

Vertebrate Any animal with a backbone. Animals without a backbone are called invertebrates.

INDEX

Scientific names

The first name of each double-barrel Latin name refers to the *Genus*, the second to the *species*. Single names not in *italic* refer to a family or sub-family and are cross-referenced to the Common name index.

94

FURTHER READING

Alexander, R. McNeill (ed)(1986), *The Encyclopedia of Animal Biology*, Facts on File, New York

Berry, R.J. and Hallam, A. (eds)(1986), *The Encyclopedia of Animal Evolution*, Facts on File, New York

Corbet, G.B. and Hill, J.E. (1980), *A World List of Mammalian Species*, British Museum and Cornell University Press, London and Ithaca, NY

Dagg, A.I., and Foster, J.B. (1976), *The Giraffe – its Biology, Behavior and Ecology*, Van Nostrand Reinhold

Groves, C.P. (1974), *Horses, Asses and Zebras*, David and Charles, Newton Abbot, England

Grzimek, B. (ed)(1972), *Grzimek's Animal Life Encyclopedia*, vols 10, 11, 12, Van Nostrand Reinhold, New York

Hall, E.R. and Kelson, K.R. (1959), *The Mammals of North America*, Ronald Press, New York

Harrison Matthews, L. (1969), *The Life of Mammals*, vols 1 and 2, Weidenfeld and Nicolson, London

Hunsaker II, D. (ed)(1977), *The Biology of Marsupials*, Academic Press, New York

Kingdon, J. (1971-82), *East African Mammals*, vols I-III, Academic Press, New York

Leuthold, W. (1977), *African Ungulates – a Comparative Review of their Ethology and Behavioural Ecology*, Springer Verlag, Berlin

Macdonald, D. (ed)(1984), *The Encyclopedia of Mammals*, Facts on File, New York

Moore, P.D. (ed)(1986), *The Encyclopedia of Animal Ecology*, Facts on File, New York

Nowak, R.M. and Paradiso, J.L. (eds)(1983) *Walker's Mammals of the World* (4th edn) 2 vols, Johns Hopkins University Press, Baltimore and London

Schaller, G.B. (1977), *Mountain Monarchs – Wild Sheep and Goats of the Himalaya*, University of Chicago Press, Chicago

Slater, P.J.B. (ed)(1986), *The Encyclopedia of Animal Behavior*, Facts on File, New York

Walther, F.R., Mungall, E.C. and Grau, G.A. (1983), *Gazelles and their Relatives – A Study in Territorial Behavior*, Noyes Publications, Park Ridge, New Jersey

Young, J.Z. (1975), *The Life of Mammals: their Anatomy and Physiology*, Oxford University Press, Oxford

ACKNOWLEDGMENTS

Picture credits

Key: *t* top *b* bottom *c* centre *l* left *r* right
Abbreviations: A Ardea. AH Andrew Henley. AN Nature, Agence Photographique. ANT Australasian Nature Transparencies. BC Bruce Coleman Ltd. GF George Frame. J Jacana. NHPA Natural History Photographic Agency. OSF Oxford Scientific Films. PEP Planet Earth Pictures. SA Survival Anglia Ltd.

6/7 A. 8 WWF/M. Boulton. 9 BC. 10/11 SA/J. Foott. 12 SA. 13 J. 14 OSF/Zig Leszczynski. 16 J. 18/19 AN. 21*tl* SAL, 21*tr* Nature Photographers, 21*b* BC. 22 Leonard Lee Rue III. 23 AH. 24/25 SA/Alan Root. 25*t* AN. 28*b* PEP/P. Scoones, 28*t* AN. 28/29 AN. 29*t* BC/H. Jungius. 31 Nature Photographers Ltd/Hugh Miles. 32 AN. 33 BC/Gerald Cubitt. 35 BC. 36/37 J. 37 AN. 38 W. Ervin/Natural Imagery. 39 J. Mackinnon. 41 Leonard Lee Rue III. 42 Leonard Lee Rue III. 43 PEP/M. Ogilvie. 44 J. 45 Leonard Lee Rue III. 46 SA/Jeff Foott. 47 BC. 48/49 PEP. 50 PEP/Franz J. Camenzind. 51 Michael Fogden. 52 NHPA. 52/53 PEP. 55 NHPA/Joe B. Blossom. 56 P. Wirtz. 57 PEP. 58/59 NHPA. 60 GF. 62/63 PEP. 64/65 M. Stanley Price. 66 M. Stanley Price. 67*t* PEP, 67*b* SA. 70 W. Ervin/Natural Imagery. 73/74 PEP. 73 GF. 75 Leonard Lee Rue III 76 BC. 76/77 Leonard Lee Rue III. 78/79 BC. 80/81*t* BC. 83 NHPA. 84 ANT. 86 J. 87 J. 89 R. Best.

Artwork credits

Key: *t* top *b* bottom *c* centre *l* left *r* right
Abbreviations: PB Priscilla Barrett

6 PB. 8 Simon Driver. 9 Jeane Colville. 11 PB. 13 PB. 14/15 PB. 17 PB. 20 PB. 23 Jeane Colville. 24 PB. 27 Denys Ovenden. 30 PB. 31 PB. 34/35 PB. 36 PB. 38/39 PB. 43 PB. 44 PB. 49 PB. 51 Jeane Colville. 53 PB. 54/55 PB. 57 PB. 59 PB. 61 PB. 65 PB. 68/69 PB. 70/71 PB. 74/75 PB. 78*b* PB. 81*b* PB. 81*c* Rob van Assen. 82/83 Dick Twinney. 84/85 Dick Twinney. 88 PB. 89 PB.